The Little Book on Learning Big Critical Thinking Skills

A Step-by-Step Guide to Developing
Critical Thinking in the Age of AI

Dr. Patricia A. Farrell, Ph.D.

ISBN: 979-8-9986832-7-5

Books by Patricia A. Farrell, Ph.D.

When You Can't Pour From an Empty Glass: CBT Skills for Exhausted Caregivers

Unfiltered Again: A behind-the-scenes look at life, healthcare, medicine and mental health

Unfiltered: Beneath the noise of our thoughts lies the true narrative of our minds.

How to Be Your Own Therapist: A Step-by-Step Guide to Taking Back Your Life

It's Not All in Your Head: Anxiety, Depression, Mood Swings & Multiple Sclerosis

Work Stress: How You Can Beat It

A Social Security Disability Psychological Claims Handbook

A Social Security Disability Psychological Claims Guidebook for Children's Benefits

The Disability Accessible US Parks in All 50 States: A Comprehensive Guide

Birding in the US NOW!: A birding guide for individuals with disabilities

DR. PATRICIA A. FARRELL, PH.D.

Sleep, Insomnia, Stress: What you don't know can hurt you

Flash Fiction

Unexpected Short Tales of Surprise

Contents

Foreword

In the modern world, there is an infinite amount of information that is often overwhelming. In a world of so much, how can we expect to navigate our small ship in this sea of not tranquility but change? Still, we must. We have no choice. Either we learn how to do what a good "pilot" does, or our ship sinks.

And we know people struggle to develop sound judgment skills. Why is that? For most of us, the struggle to develop sound judgment skills is because of changes in the world and culture. Things are not simple anymore. Today, people must handle mountains of data streams that contain artificial facts masquerading as actual information and algorithms that confirm our existing beliefs instead of challenging our thinking processes. That's the essence of the situation. We are immersed in a vast ocean of algorithms, some of which intentionally manipulate us with their biased intentions. We need to pull facts from fiction, and it will not be a simple process unless we learn to develop the skill of critical thinking now.

When I was in college, I was fortunate to read an article by Dr. Norman Geschwind. His name isn't widely known, but he was a brilliant neurologist who proposed a theory about left-handedness (Geschwind syndrome) and certain autoimmune disorders. However, that wasn't the main thing I learned from his articles.

He had a very probing mind, and his research usually started with him asking a simple question: *"Isn't it curious that...?"* The question was posed when he noticed things that seemed to connect, and then he would examine the relationship more closely to understand what it might mean. I often ask that question, and it has been a key part of my critical thinking—and it can be part of yours, too. Geschwind was looking for associations, and he didn't know whether they would be meaningful or not. But he knew he wanted to pursue them to see if there was an answer there.

While we're on the topic of Dr. Geschwind and associations, let me take this opportunity to insert one thing that may not seem necessary here. Believe me, it is. In the future, you will probably hear *people talking about correlations*. What will they say? They will express their thoughts or suggest ideas as if they have achieved an intellectual break-through. If one event occurs simultaneously with another, *it means the first event causes the second*. **No**, that's not so because **correlation can be nonsense**. I once had a professor clarify this concept for us by discussing the increased sales of ice cream in Minnesota during the summer and the corresponding rise in the incidence of rape.

A simple correlation would say that the sale of ice cream that went up was directly related to the increased incidence of rape. Could ice cream sales be responsible for this kind of criminal activity? That's nonsense, right? Yes, it is because *correlation can be meaningless*. Two things may happen, and one has nothing to do with the other. Consider a different factor, such as the weather, and compare it to the sales of ice cream and rape. What did the writer-director Alfred Hitchcock often say? He said that more murders happen in the heat of the sum-mer.

If the weather is hot, people may stay outside longer, which could lead to an increase in crime. And when the weather is hot and people feel extremely uncomfortable, there appears to be a connection between this discomfort and increased anger, as well as a diminished ability to manage emotions. It has nothing to do with ice cream. *I will almost guarantee you are going to come across that correlation argument in the future.* But you can push through it.

Remember one thing: **you don't need to know everything**, but you need to **know how to question everything**. There is no secret sauce, but there's a method that serves to enable you to *see through fantasy and find fact.*

Excessive information in our current era creates a sense of insecurity that demands top-level critical thinking skills to navigate. AI's data processing skills may surpass humans, but *it lacks critical thinking skills* (at the present time). Adapting to coding, debugging, and critical thinking keeps us current in this changing world. But there's more.

We will inevitably fall behind if we don't master this one skill. Is that what you want to do—fail and be left behind? No, you don't. You want to succeed. What's more, you want to do more than succeed. You want to stand out. Standing out isn't possible without this skill. Success comes from skill.

The future development of computer systems and educational institutions remains unclear. Artificial intelligence, certainly, has become a part of learning and the educational institutions that support it. What we know is that the educational establishment faces a dilemma regarding whether the expense and duration required for higher education justify the investment. Spending four years in training with no prospects of a positive career outcome seems like a poor choice.

In our current educational landscape, people are currently considering shortening traditional four-year degree programs to three years, and it's uncertain whether this reduction will persist or change to a simple two years in the future. *People may no longer need traditional college degrees* to secure employment opportunities or professional advancement in their chosen fields. The need for *specific certifications and training exists*, and the traditional college experience may become obsolete. Critical thinking is a skill that will remain *relevant and indispensable in all job opportunities*.

Simply put, critical thinking skills provide a solution to this challenge, which individuals *must learn on their own*. College programs previously, naturally, incorporated these skills into their curricula, but students never noticed how it affected their own intellectual development. In fact, this was what the Socratic method was all about.

The college experience included questioning (the Socratic method), but the removal of this educational element leaves us wanting. We can become victims of false beliefs and be manipulated cleverly by those who know how to control us, **who will shape our reality.** The core issue revolves around beliefs and the factors that **either support or undermine them**.

Before we go any further into critical thinking, let's make one thing clear: critical thinking is *not being negative about everything you hear* or being paranoid. Critical thinking is looking *for the truth through knowledge*. Has someone personally read the material, engaged in an interaction, or gained direct knowledge, or are they relying on someone else's opinion? If you find yourself embodying the qualities of a detective, such as Sherlock Holmes, that is entirely acceptable, as it aligns with the objective of uncovering the underlying truth in any situation. Questioning and learning how to question is the key.

This skill is the nexus of a problem, and you are trying to clarify it, get through all of the smoke and mirrors, and find the truth, if that's possible. Therefore, don't allow others to divert you from your pursuit of truth by labeling something as "fake," as this is merely *a tactic to distract you*. Throwing out words like "fake" or "false" is a tool. Don't let them distract you from the issue at hand. The specific method is often used when *you're getting too close to the truth* and poking holes in their argument. Yes, you will be up against people's egos. In addition to critical thinking, you need to inject an acceptable way of presenting everything. That's another skill you need to develop. **Quick tip here**: *Run everything through your mind several times before you allow it to come out of your mouth.* The naysayers often ascribe to the old computer adage, "Garbage in, garbage out," and they never question the garbage.

Expert forecasts predict a rise in demand for AI-assisted software, which presents an uncertain future for computer science education. Since the creation of personal computers, the market for software developers and programmers has expanded throughout each technological revolution, including the internet and smartphone eras. AI continues to progress rapidly across AI domains, and we observe new versions of current programs being launched within three to four months of their first presentation.

Truly competent AI professionals now command astronomical starting salaries, with some companies offering initial compensation ranging from $500,000 to $1 million. Imagine being a computer science major who just graduated and perhaps has one or two years of programming with some AI knowledge, and you're offered a $1M salary? That's a significant amount of money. It's also a great help with student loans.

The experts predict that professionals from different industries will develop custom industry-based programs using chatbot-style tools. New programs will operate with data sets from each field, which could trigger a technological explosion. In this new world, we need to adjust our abilities because AI capabilities surpass programming expertise. Programming education faces a diminishing role in society as universities transform their academic programs toward developing AI intellectual capabilities instead of programming skills.

What happened to Fortran, COBOL and, at my university, Snowball? Can you think of computer programs that are no longer being used today? What is the single most used programming language? The answer: Python depends on simple language commands. It is for this reason that critical thinking rises in its prominence. You will fall behind if you cannot distinguish fact from fantasy quickly and clearly and guide the AI with your ability for clarity. Make your decisions wisely, or, as they say, live to regret them.

Recent technological corporation research and artificial intelligence development efforts show machines will probably replace human operators for most programming tasks. Learning computer programming alone no longer leads to the professional advantages it used to provide. Staying ahead *necessitates critical thinking*. Welcome, traveler, to the destination that awaits you.

As I said, critical thinking is **more important than ever**, and it may be even more important tomorrow. People require independent thinking skills to evaluate evidence and *identify logical fallacies* while analyzing arguments because these abilities protect them in a complicated world. Anyone who fails to think critically can become an easy target for manipulation.

The lack of critical thinking skills makes students unable to separate fact from fiction, while unprepared professionals base their crucial choices on limited or biased information. From the research we are receiving currently, the situation will become more dire for those who *lack this skill and try to stick with computer programming*. Yes, programming may be helpful because it does lay out how algorithms are used, but it is going to diminish in professional need. This thinking skill will be an essential ability that will shape all professional domains.

Every field, profession, and corporation will require their employees to possess this skill. We see it happening now as corporations downsize and allocate jobs to those with AI proficiency. Knowing how to clearly formulate a "prompt" is another skill you are going to need to hone. No, that skill is for someone else to teach you. Start searching for a book now.

In line with applying critical thinking to our everyday lives, I've included a chapter here regarding that all-important job interview that you may be scheduling or will schedule in the future. How do you prepare for it? What do you need to do? And what's the best way to approach it?

This book delivers essential instructions for developing critical thinking skills, which remain vital in a time of unlimited questionable information availability.

A Quick Reference Guide has been added to give you fast access to decision-making procedures, and we believe it will be especially useful to you in the future. Consider making a copy and printing it for yourself for future reference. Post it near your workstation or, at home, if you are a freelancer, on the refrigerator.

BTW, If you're using this book in paperback form, I'd highly recommend that you have a highlighter ready to mark specific areas of interest. Another thing you might want to do is put small tabs into areas of the text that may relate to something important to you. Either way of noting will give you quick and easy access to something you may need to refer to in the future. It's just a small step that can prevent you from becoming frustrated.

Introduction: Why Critical Thinking Matters More Than Ever

The Day Jonathan Almost Lost Everything

Jonathan saw the email in front of him and his jaw dropped. A seemingly official message appeared on the computer screen, which displayed the bank's logo along with a sense of urgency to verify his account information or risk account suspension. How could this happen? He hadn't done anything, and yet he was in danger of having his account suspended. What was the reason? What suspicious account activity was there that might have caused this? His heart began to pound as he saw the time limitation was only 24 hours. Without hesitation, he accessed the link and provided his login information. There wasn't time to spare.

But three days later, his bank contacted him after the incident to report unexplained transactions. It began to dawn on him that he'd been a victim of a scam. Why had he panicked in the first place? It wasn't like him. During that brief moment of panic, he forgot all the critical thinking skills he'd ever learned. He had been absolutely thrown by the message and the appearance of authenticity. Not only that, but the urgent tone and financial concern activated "**System 1 thinking**," (Daniel Kahneman) which describes our automatic emotional decision-making process that frequently leads to errors. After all these years and all he'd learned about scams, they still got him. Now he was aware that everyone could be a target.

And this isn't just something hypothetical. The author of this book recently received more than one urgent text message. The message indicated that I had an outstanding traffic ticket, and if I didn't quickly go to the website and rectify it, the Motor Vehicle Department would be notified. The consequence would be the suspension of my driving privileges until I settled the fine. However, I knew that the Motor Vehicle Bureau doesn't communicate via text or e-mail messages. If I hadn't known that, I could have been roped into the scam. What's more, I have no outstanding tickets, and I know that. Checking with a colleague, I found she had also received a similar message that warned her of her driving being curtailed because of an outstanding ticket. She also has no outstanding tickets.

Another scam that I have also become aware of is associated with antivirus software, specifically McAfee and Norton. These companies are not sending out scam threats. But someone is using their names and indicating that renewals have already been processed for these services and that the individual owes $400, payable by the invoice attached. Failure to pay the invoice will result in the removal of the antivirus software from your computer. Of course, this would leave you incredibly vulnerable in our computer-dependent world. Don't fall for it, and don't act quickly.

Jonathan's experience isn't an isolated incident as I've just indicated. Millions of people encounter situations daily that seem logical to them but eventually result in financial loss. Sound judgment in decision-making depends primarily on the development of critical thinking skills by people who avoid common mental pitfalls. One reason you're reading this book is because you don't want to be victimized by scams, but you also want to learn a few things about yourself and how you can deal with any situation in your life. Let's get down to the basics.

What Is Critical Thinking, Really?

People commonly associate critical thinking with academic professors and complicated philosophical arguments. Critical thinking exists as a *fundamental practical skill* that everyone needs. Dr. Peter Facione led the most extensive critical thinking research project, which revealed that there are *six core abilities* that form the foundation for anyone's development of critical thinking skills. The six core concepts are: **interpretation, analysis, evaluation, inference, explanation, and self-regulation.**

> 1. **Interpretation** requires us to understand *the real meaning of information* rather than its surface-level presentation. A strong interpretation skill would have made Jonathan *question the bank's email method of contact*, since his bank usually communicates by mail. The email contained a specific request that needed clarification. Doesn't happen to you often? Do banks send emails that claim their services will be discontinued if we *don't respond at once*?

People receive email attachments that show **fake invoices (a favorite)** that demand payment for products they purchased or ongoing subscriptions. The company may want customers to visit their Internet site for a corrective action to be taken, or they will face disciplinary action. Going to the website often involves the process of asking you to agree to have them *take control of your desktop.* They may even ask you to *transfer money from your bank account.* The scams usually have a standard script. template, but you may not know it, and that's the reason scammers are successful. Believe me, they are very reassuring that they are trying to help you avoid an incredibly damaging incident.

2. What should you do? The process begins with an **analysis** that *separates complicated data into simpler elements* to reveal their patterns. A critical thinker would **perform three key steps** in this situation. First check the *email language, sender's address* and *previous bank communications* that you have and need to be evaluated. Banks and corporations often include this warning on their bills and invoices: "*We will never contact you by phone or through email regarding any bill irregularities.*" These warnings always explain that delinquency bills/invoices will **only be delivered by postal mail.**

3. The evaluation process requires you to *weigh the trustworthiness and reliability of your information sources.* Jonathan, if he were critically thinking, would have noted the message and the *forced time constraint* in the email. His mind would have indicated that *something seemed wrong.* We call that your gut instinct, don't we? Yes, we do.

4. You've heard of **connecting the dots** in order to see a relationship, and that is where one step of Facione's thesis lies. *Inference means drawing reasonable conclusions* based on evidence. Critical thinkers **avoid both underthinking and overthinking scenarios**. Either of these can lead to an undesired outcome. I'm reminded of what they

say in medicine: "Start low, go slow." Almost sounds like something you'd hear from a financial advisor.

5. **Explanation** is communicating your reasoning to others (and to yourself). The process of explaining the urgency to his friend would have allowed Jonathan to detect the flawed reasoning. *Speaking out loud to yourself* can sometimes provide revealing insights. You can be your coach and mentor. There's nothing wrong with talking out loud or speaking out loud to yourself, so don't think this is some type of mental disorder.

6. **Self-regulation** stands as the **highest ability** on the list because it involves tracking and improving your mental processes. This process involves **two key components**: *emotional awareness* during decision-making and *active effort to apply intentional thinking*. This is where your level of anxiety comes into play and will affect how you react in any situation requiring critical thinking. Your anxiety can serve as a beneficial emotion. But that is only if it doesn't drive you to make impulsive choices without evaluating all aspects of a situation. While anxiety can function as a motivator, it can also force people to take dangerous actions little thought. Don't overestimate your abilities and don't underestimate yourself in this situation. Remember, everything takes time and practice, and once you get that under your belt, you will have a home run.

The following series of steps can be easily remembered. The memory sentence you should keep foremost in your mind is "*I Ask Everyone Important Evidence Sources.*" It's similar to something that students of algebra will recognize. Those students know **PEMDAS** because it represents **Parentheses (or Brackets), Exponents, Multiplication and Division** (from left to right), and **Addition and Subtraction** (from left to right), which is the *order of math operations*. People use the sentence "*Please Excuse My Dear Aunt Sally*" to help them remember this concept. You can apply your new sentence right now to learn the proper sequence for critical thinking's fundamental skills. Go ahead and make up your own shortcut to remember this.

The Perfect Storm: Why Now Is Different

Modern society presents an optimal scenario for weak decision-making, according to researcher terminology. Three essential forces now combine to **make important thinking harder** while simultaneously making it more necessary than ever. The pressure to *perform excessive work leads people to our hurrying* instead of considering things carefully. And the *tendency to multitask* results in people comparing their work to others, which leads them to *believe others complete more tasks* than they do. What does it do? The belief increases our anxiety levels while diminishing our rational thinking abilities. We are sabotaging our efforts.

Every day, each of us handles about 34 gigabytes of information that would easily crash a 1990s laptop. The modern laptop, with its powerful capabilities, remains unable to manage the information load, yet we expect human brains to perform better. Our natural ability to process information *remains limited* because our brains evolved in a period when information was rare. People tend to skip detailed analysis through mental shortcuts when they face excessive data and many choices, according to research. In fact, choice can be the issue here. A very famous psychologist, Dr. Erich Fromm, once wrote a book called "Escape From Freedom." What he was discussing was that *when we have too many choices*, we become confused and anxious, and that's not a very good situation for anyone.

The Speed of Decision-Making: Digital technology demands instant responses. Don't you find yourself getting impatient when even artificial intelligence isn't giving you an immediate answer? Social media platforms use time-based metrics to measure engagement, while online shoppers need to complete purchases immediately, and news cycles operate at speeds beyond our ability to verify facts. Television news no longer turns off at 10pm but continues 24 hours a day. The artificial time pressure *pushes people to use System 1 thinking*, even though *System 2* thinking should be their choice. Here's an **overview of the two** mental processing systems.

System 1 *operates as an automatic mental system* within your brain, which functions through instinct and operates at high speed. Your automatic System 1 makes you feel uneasy around unfamiliar people. It also reaches conclusions through emotional responses, together with *first impressions and mental simplifications.* Rather than thoughtful consideration, System 1 *shoots from the hip* and immediately has a response.

On the other hand, your brain uses **System 2** thinking, which is an *intentional system*, demands analytical effort and moves at a slow pace. This system becomes active when you perform *detailed math problems*, when you *assess benefits and drawbacks* for crucial decisions, or when you *check news headlines* for potential deception.

Headlines in the media can be extremely deceptive and often engage in something called "clickbait," headlines that are meant to quickly draw you in with an emotional appeal. Often, the content of the material is not what the headline led you to believe it would be. But you were drawn in, and that's the point, especially when online. Your going to that article becomes a digit in their collection system.

The fundamental concept of critical thinking shows that System 1 makes people **accept things that seem correct** without adequate assessment. System 2 operates as the actual space for critical thinking because it requires you to **verify the truth** of the information. *"What evidence supports this? Am I missing something?"*

The goal is to determine which situations need System 2 thinking instead of attempting to suppress System 1, since your instincts remain valuable. You need to slow down and think more carefully when facing important decisions, surprising claims, and emotionally charged information. Anything with a time limit on it should demand that you sit back and give it more consideration than jumping in to meet the time limitation.

Your gut reaction functions as **System 1**, while your fact-checker operates as **System 2**.

Artificial Intelligence and Automation: While AI tools can process vast amounts of data quickly, they can also create a false sense of security. Most people now believe that computer-generated responses carry absolute accuracy. But therein lies a major fallacy. I've seen where an AI program will provide answers just to satisfy your request. Unfortunately, this doesn't mean that the answers have any validity.

AI systems use their advanced capabilities to strengthen pre-existing biases, yet they generate fake information while giving answers with confidence regarding topics they do not understand. Perhaps I shouldn't have used the word "understand" because *AI cannot understand anything*. It functions solely through programming.

If you have dealt with AI and requested something specific, have you thoroughly checked it? Are you aware that AI can and will "**hallucinate**" to provide you with information? Researchers don't believe this is done intentionally because that would mean AI has the ability to think. Computers are not sentient now.

What it is doing is fulfilling your request, as I've said, and you have to decide if it's valid or not and whether or not a website its given you actually exists. AI will *make up citations for articles* that you may be seeking. Once you go to the internet to check the validity of the article, you may get that ominous notice that it doesn't exist, it's not there, it's a **404 error.**

The Real Cost of Poor Critical Thinking

The poor application of critical thinking leads to multiple effects that surpass individual errors. The following examples involve actual scenarios.

A false news story claiming the Pope supported a presidential candidate (and even wore a trendy puffy winter coat) outpaced all actual news stories in social media distribution during 2016. Millions of people passed unverified information through social media, which could have swayed one of the most significant election outcomes in recent times.

During the initial stages of the COVID-19 pandemic, hospitals admitted patients who ingested dangerous home substances because of false medical information. Fearing that they might become ill with the virus, frightened people without critical thinking abilities for analyzing fast-moving health claims used a dangerous substance. They'd believed what they'd read on blogs on the internet, and the material was false, spreading dangerous rumors.

Businesses lose substantial amounts of money each year because employees make choices that stem from inadequate analysis and unconscious mental errors. I know of someone who was working for a major bank and was told that money was to be moved in accounts overnight. The young man, not realizing the gravity of the situation if it wasn't moved, failed to follow the order. He saw no need to do it because he did not have banking experience or question his boss about what would happen if the money wasn't moved. The next day, his boss told him they had lost over $1 million because of his failure. A tech startup lost $50 million in funding because executive team members exhibited **confirmation bias** by seeking only favorable information despite growing evidence their product was not workable. We talk at greater length about confirmation bias in Chapter 2.

But Critical Thinking Creates Success Stories, Too

One of the defining strengths of those who will advance in the sciences is the ability to question what has been accepted as truth and investigate again. This is how breakthroughs are made and people rise to the top of their professions, and it is often because of their *questioning the unquestioned*.

Strong critical thinking abilities lead to impressive professional results. Dr. Katalin Karikó worked for twenty years to develop mRNA technology even though she encountered many funding rejections and career obstacles. Personal tenacity and the ability that she knew she had were on her side. Her scientific approach to analyzing problems from different perspectives, along with that ability to assess conflicting data and maintain scientific standards, resulted in the breakthrough that enabled the COVID-19 vaccine development.

Every day, critical thinkers achieve better financial results while building stronger relationships and advancing their careers faster. These employees identify organizational issues before major emergencies occur, while parents assist children through complex social circumstances and citizens take part in thoughtful community dialogue. Critical thinking makes major contributions to people's lives and to communities. And, as we have indicated, things are changing dramatically for the future.

The New York Times published an article on July 26, 2025, about parents changing their views regarding 529 college savings plans. Because tuition costs keep rising and returns become unpredictable, these vehicles are less popular. The economic instability and questions about higher education's investment value have led some parents to postpone or stop their savings contributions. Traditional four-year degrees are no longer the only option for teens, who now consider career training and gap years as viable alternatives. Planning for careers with a college education is now in flux. Although it's usual that people may change careers three times during their lifetime, this may even change.

Identify Your Current Level of Critical Thinking Abilities

Before you develop new skills, it's important for you to understand how you currently think. Answer these questions with sincere self-reflection:

1. *When new information* opposes your existing beliefs, do you hunt for defects in the new information immediately, or do you study both perspectives with genuine interest? Most people develop *confirmation bias* because they seek information that *supports their current beliefs* while rejecting opposing evidence. Too often, it's comfortable to feel confident that you have the right answer or the right facts on your side. This is a downward spiral that will not benefit you in the future.

2. How often do you make significant decisions quickly versus the times when you take time to evaluate multiple choices? How many

choices does something that you are considering have? Complex decisions require deliberate analysis, but some choices can still benefit from rapid intuition, according to research findings.

3. During discussions about divisive matters, do you make an effort to *understand other perspectives*, or do you wait for your turn to contribute? Critical thinking manifests when someone can genuinely examine alternative perspectives.

4. Do *important people change your opinion* with additional evidence, or does your core belief set remain mostly unaltered throughout the years? *You need intellectual humility* to recognize your potential errors because it supports ongoing learning and personal development. When Columbus set out with his three ships, what did he think about the shape of the Earth? Actually, people told him he would fall off.

Your Journey Forward

This book provides a structured framework to improve your critical thinking abilities in every domain. It provides hands-on methods for evaluating information sources while teaching you to detect cognitive biases, choose under uncertain conditions, and present your reasoning effectively. Later in the book, we will also discuss how reframing is an important part of your critical thinking process. In fact, reframing is one way to help someone else accept information that they normally would not and to reconsider their own biases.

The essential benefit of critical thinking becomes clear because it lets you remain skeptical of everything while allowing you to make rapid decisions in critical situations. And critical thinking enables people to gain the skill to recognize when instinctive decisions should be trusted and when extensive evaluation is necessary.

Strong critical thinkers occasionally make mistakes, yet this skill remains the goal of critical thinking training. The development of dependable thinking/questioning systems will enable you to maximize correct decisions when facing crucial situations.

In our information-rich society, the most important skill that you can have in this AI-enabled world is knowing how to handle data rather than merely accessing it. This process of developing wisdom starts with critical thinking, and critical thinking begins when we choose to examine our thinking processes more thoroughly. You may be a bit hesitant, but that's normal because all of us would rather not experience the discomfort of knowing that we have some area where we lack ability. But how do you learn if you don't seek to find where you may be deficient? I shouldn't say "deficient" because that almost portends failure, and I don't believe that, in learning especially, that failure is an option. Each time you "fail," what you are actually doing is learning something.

Chapter 1: Understanding Your Thinking Mind

Marcus held a strong belief in logic as his fundamental principle of life. As an engineer by profession, Marcus used to approach problems with a methodical approach and allow data to steer his choice of action. When he began to look at how he had denied anything other than data to make decisions, a lightbulb went off.

He felt stunned when he discovered he had been taking a different route to work for months because of a disturbing accident he witnessed at that location. A single powerful memory had taken over his logical thinking process. How often has an upsetting memory caused us to make decisions affected by it?

Psychologists have known for decades that every person possesses two separate thinking systems that operate independently of each other. These systems' knowledge serves as the base for better decision-making across every life choice, starting from career choices and extending to shopping habits.

Your Two-System Brain

Daniel Kahneman, who received the Nobel Prize, brought a paradigm shift to our thought process by establishing two mental processes, which he labels **System 1 and System 2**. These systems *operate as thinking modes* rather than physical brain regions that control every decision you ever make. Yes, our brain is divided into two hemispheres, but what we are talking about here is thinking, not physical divisions in the brain.

System 1: Your Fast Brain

It operates automatically *through intuitive processes* that work at high speed. The system enables you to understand these written words right now with no need to decode them letter by letter. The system instantly detects facial expressions as well as recognizes threats while delivering rapid assessments. In effect, it's a survival system that alerts us when instant action is required to save us from something or someone.

Real-world System 1 in action:

• **Shopping**: When entering Target for toothpaste, you unintentionally end up spending $87 on unplanned purchases

• **Social media**: Without finishing the article, you automatically share a headline that triggers your anger

• **Driving**: Your mind automatically follows your regular home path even when you expected the construction along the way

• **First dates**: You reach a decision about interest within a few short minutes based on things that remain unclear to you

• **Job interviews**: Selection decisions made by hiring managers *during the first thirty seconds* lead them to spend the rest of the interview verifying their initial assessment

This system developed to **keep us alive**. Our ancestors who delayed thinking about situations became predators' food. The ones who sensed fear right away prepared to run and survived to become our ancestors.

System 2: Your Slow Brain

System 2 thinking requires *deliberate analysis* that involves effort and is analytical. Thinking occurs when you actively consider how these concepts relate to your life instead of simply reading them.

Real-world System 2 in action:

• **Major purchases**: Researching cars for weeks, comparing features, reading reviews, calculating financing options

• **Tax preparation**: Working through forms step-by-step, double-checking calculations

• **Learning new skills**: The first time you practice guitar chords or make a complicated recipe requires focused attention.

• **Workplace decisions**: Analyzing quarterly reports to determine budget allocations

• **Relationship conflicts**: During arguments, you need to *distance yourself from the moment* to see things from your partner's point of view before you reply

System 2 represents a strong mental system, yet it requires significant mental energy to function slowly, which makes us seek to avoid it during times of fatigue or stress even though these are *precisely the situations when we need it most.*

We must recognize that both System 1 and System 2 will remain active components of our existence. Occasionally, we utilize both systems. Depending on which system you use, it can produce costly problems in modern life, so you should think carefully before moving forward.

System 1 Traps in Action

1, The *Availability Bias* Shopping Mistake:

The news coverage of a plane crash caused Jennifer to abstain from flying for two years while she drove thousands of miles for business travel. The statistics showed she faced a 150 times greater risk of death during car trips than the flights she avoided taking. *Her System 1 brain labeled flying as unsafe* because plane crashes were *more memorable* than car accidents that rarely received local news coverage. She was, in effect, a victim of media selective coverage.

2. The *Confirmation Bias* Investment Error:

David owned a specific EV vehicle while reading only positive articles about electric vehicles and *ignoring negative reports* that he dismissed as "oil company propaganda." During a stock drop, David lost $15,000 because he did not take any risks that could have *challenged his positive perspective*. Have you ever found yourself, ignoring certain reports or issues because you found them contrary to what you currently believe? If so, you've been practicing the confirmation bias.

3. The *Halo Effect* Hiring Mistake:

During the hiring process, Sarah's team selected a charismatic candidate who performed outstandingly during interviews and held an MBA from a well-known university. After six months of employment, the team discovered that the candidate had fabricated his credentials and lacked essential technical abilities required for the job. The positive first impression of the candidate created a "halo" effect that *stopped the hiring team from detecting warning signs* that could have been observed in alternative candidates. I've seen this in research settings where individuals come in, seem to have the appropriate skills to perform the task, and then, once they are hired, the team discovers the skills they said they possessed are absent. Not only did the research project lose valuable time, but it also negatively impacted the group's morale.

System 2 Traps in Action

1. The *Analysis Paralysis* House Hunt:

During their house search, Mike and Lisa devoted eighteen months to assessing all neighborhoods and school districts, as well as market trends, together. Their 18-month home search became a $200 monthly cost increase when interest rates rose by 2% during that period, and they spent thirty years with this additional expense. Think how much they were going to pay over the life of that mortgage because of their inability to stop making judgments that, in the long run, proved useless.

2. The *Motivated Reasoning* Divorce:

Tom spent hours searching for legal backing to support his position, but refused to consider evidence backing his wife's point during their divorce proceedings. His analytical skills helped him construct evidence that supported his desired opinions rather than revealing actual facts, which resulted in costly legal bills.

3. Developing *Thinking Awareness*: The Metacognitive Edge

The process of System 1 elimination remains impossible together with the use of System 2. Metacognition involves understanding the operational processes of your brain. Your brain operates with an automatic control system that helps you select the most suitable mental tools based on the current situation. It also has that other system that helps to somewhat control the first system when making decisions.

You have been operating, potentially, in one manner for quite a number of years, but you probably never realized that you were *under the control of habit*—a habit that can be eliminated through studying alternative ways of thinking and behavior modification. The vital idea of change is important yet demands effort, and you must integrate the items from this book *into your daily activities*. Your life will transform positively, and you will achieve better outcomes while experiencing increased satisfaction.

Week 1 Exercise: Thinking Detective

Note three different mental states that you experience during the next seven days. You want a record of your observations so you don't forget them. Yes, you will encounter all of these states during this seven-day period:

1. Automatic reactions: Strong immediate responses to information or situations

2. Deliberate analysis: Times you consciously worked through a problem

3. Confidence gaps: Moments you felt certain but were wrong, or uncertain but were right

***Real examples from life*:**

A Facebook post exposed the uselessness of taking multivitamins for me. This information immediately triggered questions about my multivitamin use, and what was the result? My defensive System 1 confirmation bias response kicked in. The study had examined a single vitamin while ignoring all other multivitamins (I should have begun with System 2). The research studied vitamins but failed to investigate the specific vitamins I was taking. Too often, in research, we fail to look at the finer points of the studies that are proclaiming "breakthrough" results, and we are misled.

During a meeting, our boss sought my opinion regarding new software. My evaluation of the software led me to express uncertainty. Major security issues existed in the software at the time. I initially doubted its use, which later proved my System 1 intuition correct.

I spent twenty minutes selecting which $4 coffee to purchase at the new cafe. My decision-making process had become too detailed for a decision that required no complexity, so I recognized *System 2 overkill in my thinking*.

Spotting Your Personal Bias Patterns: Practical Recognition

The identification of biases in real-world situations improves through specific bias understanding. Our personal biases operate in our daily lives, but many times we remain unaware that we display biased thinking. We usually hide our biases from ourselves, but this system helps us notice them, which proves to be very valuable to us.

Anchoring Bias in Daily Life

Real estate agents use this bias. The first house you view during your house search establishes your basis for assessing every house that follows it. The $500,000 starter home looks reasonable after inspecting the $750,000 house but costs more than the $300,000 fixer-upper.

The initial pay amount set by the boss determines what workers think about following salary offers. When your initial expectation is 8%, you will react negatively to a 4% offer.

You must *establish reference points* before making accurate assessments. When buying a car, research vehicle prices extensively before visiting automobile dealerships.

Confirmation Bias in Information Consumption

Media: The news sources that supported Alex's political beliefs *became his exclusive news outlet*. During the 2020 election, Alex experienced shock about the results because he had unconsciously blocked opposing information for many months.

Health Example: Maria thought that eliminating gluten from her diet would address her energy concerns. The pattern of her good health days following gluten elimination became obvious to her, but she chose to ignore the days she felt tired even when she followed the diet exactly. It took her several months to discover that *her energy levels were linked to sleep quality* instead of food choices.

Before researching something that matters to you, note down evidence that might change your perspective and actively search for that evidence.

Availability Bias in Risk Assessment

James became sure that a housing market collapse was imminent after watching a documentary film, so he kept his savings in cash for three years but missed out on significant investment gains. Housing market crashes appeared more realistic in news stories because they used vivid images, even though actual statistical data showed otherwise. Again, this is an example of how a visual can be used to sway your opinion.

Susan took up driving her children to school instead of walking them there after becoming aware of a local child kidnapping incident. The odds of their safe journey by car, in her mind, exceeded their risk of being abducted while walking. Statistics, however, have shown that car accidents are much more frequent than children being kidnapped while walking to school.

Risk evaluation demands you ask whether *the thought is based on common experiences or vivid memories*.

Three Daily Metacognitive Habits That Work

1. The $50 *Pause* Practice

Before spending more than $50 on a purchase or posting news on social media or forming first impressions of people, you should ask which of your thinking systems is required here.

During her work lunch break, Janet almost bought a $200 exercise bike after seeing the product in an Internet video. She *paused* to recognize her *emotional impulse to buy* the exercise bike. What caused her to think she might need this bike? Janet had not completed her daily workout and that spurred this impulse. But rather than immediately purchase the unit, *she waited for seven days* until she realized she didn't need any home exercise equipment.

2. The *Alternative Explanation* Game

Create two additional explanations for every irritating situation beyond your first impression. This may not always be easy, but it's good practice to give your brain that exercise.

Real life examples:

• A colleague missed an important meeting. First thought: "He doesn't respect our time." Alternatives: "Emergency at home," or "Double-booked and didn't know how to handle it."

• Friend hasn't returned texts for three days. First thought: "She's mad at me." Alternatives: "Phone broke," or "Dealing with family crisis."

• A driver cuts you off. First thought: "Aggressive jerk." Alternatives: "Didn't see me," or "Rushing to hospital emergency."

3. The *Evidence Question*

Each day, you need to test your beliefs by asking, "*How do I know this statement is true*?" Follow the trail of beliefs back to their original sources. You may be surprised how you came to these beliefs and even wonder how you could have been convinced so easily.

The actual case of Tom revealed he held the belief "*breakfast is the most important meal*" without understanding its original source. The creation of his belief about breakfast came from 1960s marketing campaigns by cereal producers rather than actual scientific nutritional findings. How often have you asked, "Who underwrote the research for this product?"

We often see this belief in commercials for several health or sleep aids or even skin creams. How many hospitals are given a large quantity of free samples of a particular over-the-counter pain medication that is then distributed to patients being discharged? How many people were in the sample that tested a face cream and found it reduced wrinkles? Where was the research conducted? Not only who ran the research, but who published it and who might own the publishing company is also important. Are you aware that at one time, *cigarette commercials on TV claimed that more physicians smoked* a particular brand than any other? Of course, the research on cancer and smoking has eliminated that belief.

Building Bias Resistance: Practical Exercises

Exercise: Belief Archaeology

This exercise requires selecting *three fundamental beliefs* that determine your *major financial, relational, professional and wellness decisions*. For each, identify:

Belief: "Renting is throwing money away."

- Original source: Parents always said this during childhood
- Supporting evidence: Building equity vs. paying a landlord
- Last questioned: Never seriously examined alternatives

• What would change your mind? Discovering the total cost of ownership, including maintenance, taxes, and the cost of the down payment.

Belief: "Natural products are always safer."

• Original source: Health blog I read five years ago.

• Supporting evidence: Processed foods have caused health problems

• Last questioned: When a friend got sick from "natural" supplement

• What would change your mind? Learning about natural toxins and rigorous testing of synthetic products.

Exercise: Decision Speed Audit

One-week assessments of major decisions should be performed *using either "Fast" (instinctive) or "Slow" (logical)* categories.

Fast decisions that *worked*:

The restaurant warning sign prevented me from eating there before discovering its health violations.

The friendly real estate agent received my selection over the pushy agent.

Fast decisions that *backfired*:

A friend's stock recommendation about trending investments resulted in a loss of $500 for me.

The decision to take on additional work because my boss appeared stressed ended in burnout.

Slow decisions that *worked*:

I conducted thorough research to select appropriate contractors for my home renovation project.

Carefully evaluating job offers, including long-term career impact.

Slow decisions that were *overkill*:

The selection of a casual dinner restaurant took three hours in total. Three hours?

I studied 20 review assessments prior to buying a $15 phone cover. Not time well spent.

A tennis enthusiast friend of mine enjoyed playing tennis thanks to his luck in having semi-professional tennis friends who played the sport. He developed an intense fixation on how the professionals modified their rackets to achieve better performance during matches. He asked everyone in his path about the correct stringing configuration for rackets and string tension levels.

The staff at a specialized store received detailed instructions from him after he visited the shop. The selection process for his racket *took several weeks* yet failed to improve his game. He followed the same pattern in purchasing a convertible sofa, just like with tennis rackets. The sofa project took two months. Was he satisfied with either question? No, he still thought he should have given both of them more thought and time. You couldn't change his mind.

Red Flag: When to Override System 1

Right now, you should understand how to detect particular situations that could lead your fast brain to provide wrong information.

Financial Red Flag:

- "Limited-time offer" that creates urgency
- Investment opportunities that seem too good to be true
- Purchases made while emotional (angry, excited, sad)

Relationship Red Flag:

- Strong instant attraction or dislike of someone important
- Feeling pressured to make quick commitments
- Relationship decisions made during conflict

Professional Red Flag:

• Job offers with pressure to decide immediately ("Decide now, or the offer is off the table.")

• Business decisions based on one dramatic success or failure story

• Hiring based primarily on "gut feeling" about candidates. One museum hired an attractive young man to be in charge of their computer setup. He promptly bought new equipment for himself and arranged a trip to Europe to look at museums there. Once he had the new equipment and returned from the trip, they found his skills were seriously lacking, and they'd made a mistake in hiring him.

• When news reports match exactly with your preexisting opinions

• A health expert provides basic answers to complicated health issues. Unfortunately, health can be quite complex, and simplistic answers, while attractive to the media, may fail in supplying patients with needed information. My experience in television media has led me to see quite a bit of this, where a person is telegenic but lacking in credible information.

• Political information depicts its opponents as completely irrational. At this point, we all know about smear campaigns and how political consultants use them to gain votes for their candidate.

Chapter 1 Skills Checkpoint

You should be able to do the following before moving forward:

___ Identify your thinking systems: Recognize when you're using System 1 vs. System 2 in actual situations

___Spot personal biases: Catch yourself making biased judgments as they happen.

___Practice metacognition: Regularly question your own thinking processes.

__Apply the pause practice: Know when to slow down for more careful analysis.

Self-Check Questions:

1. Detail a System 1 decision from previous experiences that you can describe and that involved a particular situation where its appropriateness needs assessment.

2. Daily decisions are most often influenced by which cognitive bias? Give a specific example.

3. This week I choose to establish one metacognitive habit while describing my strategy for maintaining its practice.

Real-World Application Challenge:

You must determine the appropriate thinking system to use before making any choice that costs more than $20 or requires more than 20 minutes of your time for the upcoming three days. *Observe how your decision quality changes.*

The search for perfect thinking remains impossible, because thorough analysis can cause total mental paralysis. The goal is to detect incorrect, automatic mental outputs and determine when you need more time for extra evaluation.

You are *developing foundational knowledge* to handle the overwhelming digital information that defines our modern world. These skills don't happen overnight or just by reading one book. You need to keep yourself aware of how you are acting, do a bit of analysis regarding which mental system you are using, and then proceed with appropriate action.

Chapter 2: Confirmation Bias: When We See Only What We Want to See

Before laying out what is needed in terms of critical thinking, let's consider one major issue that will confront all of us: confirmation bias. Although we sometimes blind ourselves to our biases, it is imperative that we permit ourselves to carefully consider them. Often, it's helpful to know that we are not alone and that major corporate and political executives, as well as those in entertainment, healthcare, and other areas, have made disastrous mistakes. They failed to do what was needed to analyze what was happening, and they folded to their confirmation bias. **What exactly is confirmation bias?**

Confirmation bias is one of the most pervasive and dangerous obstacles to clear thinking. It's the *tendency to search for, interpret, and remember information in ways that confirm our pre-existing beliefs* while giving disproportionately less consideration to alternative possibilities. Unlike deliberate deception or conscious manipulation, confirmation bias often *operates below our awareness*, making it particularly insidious. You may not know your biases, but you'll learn about them.

One area where most of us have absolutely no knowledge or awareness of it is when people are writing computer programs. Bias is inserted into the program algorithm, and then it is used for future algorithms. All of these biases, which the programmer is unaware of, are perpetually pushed into other algorithms. The danger stays hidden from view.

Think of confirmation bias as *wearing rose-colored glasses* that filter out certain colors while enhancing others. When we're strongly committed to a particular belief or idea, our minds unconsciously seek out information that supports that belief while dismissing or overlooking contradictory evidence. This mental filtering happens so automatically that we often don't realize we're doing it. In essence, it is almost a matter of self-esteem and no one wants their self-esteem damaged by contradictory information.

When is this type of bias most concerning? Confirmation bias becomes especially dangerous when *combined with positions of power and responsibility*. Corporate leaders have all demonstrated this bias in ways that ultimately led to catastrophic failures. By analyzing their specific examples, we can better understand how confirmation bias operates and learn to recognize it in our thinking.

Elizabeth Holmes: Filtering Out Scientific Reality

Elizabeth Holmes's story provides perhaps the clearest example of confirmation bias in action. Medical research has been searching for the ultimate technological advancement in the analysis of specimens, particularly body fluids such as blood. Holmes had a corporation, Theranos, that indicated a small unit they were manufacturing could provide an inordinate number of tests with a simple small sample of blood.

Theranos claimed their "Edison" device could *run hundreds of different blood tests* using just a tiny amount of blood—sometimes as little as a single drop from a finger prick. The business claimed to conduct over 240 tests, ranging from identifying drugs like cocaine or marijuana in your system to screening for conditions like diabetes and cancer. They said this could all be done quickly and accurately using their compact, automated machines.

Traditional blood tests usually need blood drawn through a large needle, which can be painful and expensive. Theranos promised to change all that. They believed their Edison device could replace painful needle draws with that simple finger prick and test for hundreds of diseases just as accurately as regular blood tests. If the machine could do what they said, it was an incredible revolutionary breakthrough in medical technology.

Later reports in the media as well as at legal trials would indicate the reality was very different. The Edison machine had many difficulties—pieces would fall off, doors wouldn't close, and it couldn't even control its temperature properly. Only one test—for the herpes virus—was ever approved as reliable by the FDA. Most of their tests were actually done using regular machines from other companies, not their revolutionary Edison device. How could this happen, and where was the bias?

The Bias in Action: When Theranos scientists and engineers allegedly repeatedly told Holmes that the technology wasn't working as promised, she consistently dismissed their concerns rather than treating them as valuable data points. Former employees reported that Holmes would question the competence of scientists who raised technical objections rather than seriously considering whether their concerns might be valid. As so many who don't recognize their bias do, she was convinced the machine was workable and that the scientists were incorrect.

For example, when lab workers pointed out that the tiny blood samples weren't sufficient for accurate testing, Holmes didn't treat this as important feedback about a fundamental flaw in her approach. Instead, she seemed to view it as a temporary obstacle that could be overcome through determination and willpower. Workarounds were sought, according to sources.

The Filtering Mechanism: Holmes appeared to interpret all information through the lens of her pre-existing belief that the technology could work. When presented with positive results or encouraging developments, she gave them full weight and attention. When presented with negative results or technical challenges, she either dismissed them as temporary setbacks or questioned the competence of the people raising concerns. Everything was seen through the lens of her personal bias.

This *selective attention* extended to external validation as well. Holmes publicized positive media coverage and celebrity endorsements while dismissing criticism from medical experts and scientists as coming from people who "didn't understand" the revolutionary nature of what Theranos was attempting.

The board of directors of Theranos included *Channing Robertson* (Stanford professor and the first board member), *George Shultz* (former U.S. Secretary of State), *William Perry* (former U.S. Secretary of Defense), and *Dr. Henry Kissinger* (former U.S. Secretary of State). Other members included *Sam Nunn* (former U.S. Senator), *Richard M. Kovacevich* (former Wells Fargo CEO), and *Fabrizio Bonanni* (former Amgen executive). In fact, the board was more than stellar in its high-profile membership. Anyone with any hesitation about the company could look at the board and feel an incredible measure of confidence in the machine.

The Consequences: This confirmation bias prevented Holmes from making necessary course corrections that might have led to a more realistic and achievable approach to blood testing innovation. Instead of acknowledging fundamental technical challenges and adjusting her timeline and claims accordingly, she doubled down on promises that couldn't be kept, ultimately leading to the complete collapse of the company.

Adam Neumann: Redefining Reality to Fit the Vision

Adam Neumann's confirmation bias manifested in his persistent *reframing of WeWork as a technology company* rather than a real estate business. This wasn't just a marketing strategy—Neumann appeared to genuinely believe that WeWork was fundamentally different from traditional real estate companies, despite seemingly overwhelming evidence to the contrary.

The Bias in Action: When financial analysts and real estate experts pointed out that WeWork's business model had fundamental vulnerabilities—particularly the mismatch between long-term lease obligations and short-term customer commitments—Neumann consistently dismissed these concerns as "old-fashioned thinking" that failed to understand WeWork's revolutionary approach.

Neumann seemed to be selectively *focused on metrics that supported his vision* while downplaying or ignoring metrics that revealed problems. He emphasized growth in membership and revenue while minimizing the significance of massive losses and unsustainable unit economics. When experts pointed out that WeWork's customers were primarily small businesses and freelancers who would be vulnerable in an economic downturn, Neumann was unmoved by these concerns. He argued WeWork was creating a new category of business that would not be subject to traditional economic cycles.

The Filtering Mechanism: Neumann's confirmation bias operated through *his redefinition of WeWork's fundamental nature*. By insisting that WeWork was a technology company focused on community and collaboration rather than a real estate company based on traditional real estate metrics that he saw as irrelevant. For him, it was their inability to look into the future of business and realize that he had a prize under his control.

This reframing allowed him to interpret all information through *the lens of his preferred narrative*. It must have been frustrating for him, undoubtedly. The problem lay in his rapid expansion, which he believed would be sustained by investor money. But rapid expansion and sustainable growth fueled by investor money were an illusion. Massive losses weren't a sign of fundamental business model problems—they were *necessary investments* in building a transformative company, so he decided. Rather than see the reality of losses, meaning his vision had a fault in it, he proceeded.

The Consequences: Neumann's confirmation bias prevented him from recognizing and addressing the genuine risks in WeWork's business model. When the IPO process forced detailed financial disclosure, the disconnect between Neumann's narrative and the company's reality became impossible to ignore, *leading to his ouster and the company's dramatic devaluation.*

Sepp Blatter: Rationalizing Corruption as Politics

Sepp Blatter's confirmation bias operated through *his interpretation of corruption allegations as politically motivated attacks* rather than legitimate concerns about FIFA's (Fédération Internationale de Football Association) governance. His deep belief in his mission to democratize world soccer and reduce European dominance created a mental framework that *filtered out uncomfortable truths* about the organization he led.

The Bias in Action: When journalists, prosecutors, and reform advocates raised concerns about corruption within FIFA, Blatter consistently interpreted these concerns as attacks from European interests that were upset about losing their traditional control over world soccer. This interpretation may not have been entirely wrong—there were political and cultural elements to the criticism Blatter faced. However, his focus on the political motivations of his critics prevented him from seriously considering whether their substantive concerns about corruption might be valid. He was a man on a mission to turn soccer around, and he failed to see that there were other challenges to be faced.

Blatter appeared to selectively attend to information that supported his narrative of FIFA as a democratizing force in world soccer while dismissing or minimizing information about systematic corruption. He emphasized FIFA's financial growth, the expansion of World Cups to new continents, and investments in soccer development while downplaying persistent allegations of bribery and fraud.

The Filtering Mechanism: Blatter's confirmation bias operated through his broader narrative about FIFA's mission and his role in advancing that mission. Because he genuinely believed he was fighting for important principles—global soccer development and reduced European dominance—he could rationalize questionable practices as necessary political compromises.

This created a mental framework where criticism was automatically interpreted as politically motivated rather than substantively valid. When officials were accused of taking bribes, Blatter could dismiss these allegations as attacks on FIFA's independence rather than serious governance problems that needed to be addressed.

The Consequences: Blatter's confirmation bias prevented him from recognizing and addressing the systematic corruption that was undermining FIFA's credibility and mission. By the time the extent of the corruption became undeniable, the damage to FIFA's reputation was severe, and Blatter's own career was destroyed.

The Downfall of BlackBerry: When Success Breeds Blindness

BlackBerry's failure *stems from its success*, which made leaders unable to see their mistakes.

Mike Lazaridis presented the ideal combination for leading the smartphone revolution. Through his role as co-founder and co-CEO at Research In Motion, he established BlackBerry as the leading mobile communication platform trusted by executives and government officials and ordinary users who needed physical keyboards for typing and instant messaging.

During the company's success period, the executive displayed rising confidence levels. Lazaridis had every reason to feel confident. The business market belonged to BlackBerry, while the stock price experienced remarkable growth from $10 in 2003 to reach $140 in 2008. Users recognized the devices for their secure and dependable nature and enjoyed the sound of typing on physical buttons. Smartphone users used "BlackBerry" as a synonym for mobile phones in the same way people use "Kleenex" for tissues, Frigidaire was for a refrigerator, and Xeroxing for making copies. It was a time when the lexicon was rapidly changing.

The company's success fostered an unhealthy attitude, leading employees to believe they understood their customers better than anyone else. Their corporate dominance stemmed from creating devices for serious business users who required keyboards for efficient typing and enterprise-level security features.

The iPhone introduction marked a threat that transformed everything

Steve Jobs presented the iPhone to the public in January 2007. Lazaridis failed to appreciate this innovative product during its unveiling. The executive team focused on the iPhone's limitations instead of its benefits. Business users would never switch to glass-based typing, according to the company's belief. Plus, the battery life of this device remained shorter than that of BlackBerry devices and the system lacked the security features that corporations needed for their operations.

Lazaridis publicly debated whether people would actually use their phones to access internet content or watch videos through small displays. Then Lazaridis deliberately collected evidence that supported his preconceived notions about smartphones yet failed to recognize the major changes in consumer preferences. Who was behind the business communications curve here?

Confirmation Bias in Action

Lazaridis ignored multiple warning indicators while providing alternative explanations. The rapid increase in iPhones led him to believe that Apple used effective marketing techniques instead of actual consumer interest. RIM's CEO believed serious business needs required BlackBerry devices because he observed junior employees *using iPhones personally* while working with BlackBerrys at the office. Did the personal use of iPhones distract him from critically questioning his own product?

Lazaridis *selected executives who shared his belief* that smartphones should prioritize keyboards and business operations. The leadership team doubted the market research findings about consumer interest in touchscreen devices and multimedia features instead of reassessing their business strategy.

The company culture inside the organization reinforced these fundamental beliefs. The development of touchscreen devices and app stores was redirected towards projects that built upon BlackBerry's traditional capabilities by engineers who suggested these innovations. He believed his team understood what worked best, and their methods remained as established business practices.

The Inevitable Collapse

The decline of BlackBerry became apparent to everyone during 2010. After the iPhone launched, it created a complete app-based ecosystem that turned smartphones into essential tools for entertainment and business purposes. The market experienced a flood of Android devices, which duplicated BlackBerry features at different price levels.

The company launched delayed and insufficient responses to its market challenges. The touchscreen devices from BlackBerry felt inferior to rival products in the market. Their application store lacked both variety and excitement. After establishing itself as the leading mobile communication company, the organization became recognized as outdated and obsolete in the market.

The Human Cost

The consequences extended beyond market losses and stock value changes. Thousands of workers became unemployed when Black-Berry started terminating employees and shutting down facilities. People who placed their trust in the company leadership lost their entire investment value. The economic situation of BlackBerry's headquarters community, together with its surrounding areas, suffered a major decline.

Lessons for Future Leaders

The case of Lazaridis demonstrates how successful leaders can become blind to reality because of their confirmation bias. His previous achievements and domain expertise created obstacles that prevented him from perceiving new market trends. The self-assurance that developed at BlackBerry proved fatal because it blocked the team from objectively assessing new data.

The situation illustrates that fast-moving industries cause established advantages to transform into potential vulnerabilities unless leaders display a willingness to question their beliefs while embracing challenging realities.

The Political Blindness: When Believing Your Own Hype Backfires

Eric Cantor dominated all aspects of politics in Virginia. His role as House Majority Leader, along with his second-ranking position in Congress, helped him build a strong political network over seven terms. Many observers saw Eric Cantor as a future Speaker of the House because he controlled significant funds and was well known across the country, all while maintaining a secure Republican district.

The Bubble of Certainty

The campaign team relied on experienced political professionals who had repeatedly achieved electoral victories. They polled the electorate to confirm Cantor's powerful position against economics professor Dave Brat in the primary race. The campaign used internal polling data, which showed Cantor leading by double digits, to dismiss the primary challenge as insignificant.

The campaign also concentrated on building national relationships and fundraising for future leadership battles rather than engaging with local voters through grassroots activities within the district. Campaign leaders dismissed warnings from local activists about growing conservative voter discontent toward Cantor, viewing these concerns as stemming from an extreme minority.

Warning Signs Ignored

Several warning signs emerged that the campaign either downplayed or rationalized. Town hall meetings revealed increasing hostility from voters upset about immigration policies and Cantor's detached relationship with local community issues. The campaign staff ignored the genuine threat of voter anger, believing it was driven by organized groups unlikely to produce meaningful election results.

Cantor's team dismissed the influence of talk radio stations and grassroots conservative media outlets because these groups were actively supporting Brat. They believed traditional campaign resources—including financial backing, official support, and professional organization—would outperform Brat's straightforward but targeted campaign approach.

The Shocking Reality

Cantor faced an unexpected defeat on primary night, becoming one of the biggest losers in recent political history after losing by 11 percentage points. The campaign's assumptions about money and establishment backing proved incorrect because Republican primary voters favored an outsider over an insider.

This case illustrates that politicians who rely heavily on *data supporting their preconceived ideas* can become disconnected from their core supporters.

But a similar presidential election campaign also suffered devastating results, and that was in the election of 1948.

After World War II, President Harry Truman looked doomed. The country faced tough economic problems, and many thought

Truman—who only became president when Roosevelt died—wasn't strong enough to handle it. His own Democratic Party was splitting apart. Southern Democrats broke away to form the Dixiecrats with Strom Thurmond, while Henry Wallace ran as a Progressive. The Democratic vote was split three ways.

Everyone expected Republican Thomas Dewey to win easily. Dewey played it safe, but Truman had other plans. He jumped on a train for his famous "whistle-stop" campaign, traveling across America and stopping in small towns to blast the Republican Congress and promise to fight for working folks. Truman was not about to give in to the confirmation bias of the Republican Party.

On election night, everyone got the shock of their lives. Truman pulled off one of the biggest upsets in American history, crushing Dewey 303 to 189 in the Electoral College. It demonstrated that an underdog with unwavering determination can overcome all obstacles. *But what was the one factor that many people missed* in the polling? The polls were of people who *owned telephones in their homes*, and that was specifically Republican voters. Democrats had few voters with phones in their homes. The assumption was to bolster the Republican polls falsely, even to the leaders, and lead to this loss.

The Space Race Stumble: When Pride Prevents Progress

Dennis Muilenburg seemed to be the ideal executive for Boeing when the competition was intense in the aviation industry. During his tenure as CEO from 2015 to 2019, he took over a firm that maintained outstanding excellence in safety and engineering. The Boeing Company had been the undisputed leader of commercial aviation for many years, while the 737 series emerged as one of the most successful aircraft families of all time. Muilenberg expressed certainty that Boeing would produce an updated version of the successful 737 following Airbus' introduction of the A320neo fuel-efficient aircraft.

The Pressure to Compete

The pressure from Airbus competition forced Muilenburg to speed up Boeing's response. Instead of building a brand-new aircraft, Boeing chose to modify their established 737 with bigger and more efficient engines. The airline preference for the 737, combined with existing pilot training expertise and Boeing's quick-to-market capability, made this approach appear sensible.

But there was one additional issue: the larger engines created a problem. The 737's low-sitting frame required a different engine installation, which led to changes in the aircraft's handling properties. The solution Boeing created for the 737 was a new software system known as MCAS (Maneuvering Characteristics Augmentation System), which would automatically adjust the plane's angle to prevent stalling.

Confirmation Bias Takes Hold

The leaders, including Muilenburg, strongly believed their engineering solution was correct while maintaining Boeing's reputation for safety was unassailable. During the early development of MCAS, the leadership focused on positive data regarding system performance rather than studying potential failure scenarios.

The company culture at Boeing supported the belief that its methods were sound. And the Boeing leadership frequently silenced engineers who voiced concerns about pilot training needs and system backup requirements by asserting their traditional engineering excellence would dominate.

Muilenburg presented Boeing's safety track record to the public while dismissing accusations that Boeing sacrificed safety to outdo Airbus. In addition, Muilenburg chose executives who shared his confidence in Boeing's engineering judgment being superior to external criticism.

The Tragic Reality

The 737 MAX crashed within five months, resulting in the deaths of 346 people, before demonstrating dangerous MCAS system failures. The investigation revealed that pilots did not receive sufficient information about the system, while the software depended on data from a single faulty sensor.

The crisis led to a global 737 MAX aircraft ban, which cost Boeing more than $20 billion while destroying its reputation as a safety-focused company. CEO Muilenburg eventually lost his position, and Boeing was involved in criminal charges for deceiving regulatory bodies.

The Human Cost

The fatal accident resulted in the deaths of numerous people and created massive job losses for Boeing staff members. Airlines suffered enormous financial losses because their aircraft remained grounded, while public confidence in aviation safety suffered a major decline.

Leaders who demonstrate excessive confidence in their expertise while avoiding critical evaluation of their assumptions can lead to deadly outcomes, as demonstrated by this case.

The AI Hype Trap: When Smart People Make Dumb Bets

OpenAI made Sam Altman its public face when he took control of the organization in 2019. The non-profit research organization OpenAI under Sam Altman's leadership seemed destined to guide the responsible creation of artificial general intelligence until his departure. His choices about commercial partnerships demonstrated how confirmation bias affects well-intentioned leaders even when they make decisions.

The Pressure to Scale

OpenAI's survival required massive funding for AI research, according to Altman, while he negotiated the Microsoft partnership because Google and Facebook stood as major competitors in the market. Altman presented the two major investments from Microsoft as crucial measures to achieve OpenAI's mission objectives after the company received its first billion-dollar investment in 2019 and its subsequent larger partnership in 2023.

Further, Altman fully endorsed the belief that major corporate collaboration represented the exclusive way to develop responsible AI at scale. He ignored the warnings from both AI researchers and ethicists, who feared Microsoft's profit-driven goals would challenge OpenAI's initial safety-oriented purpose.

Warning Signs Dismissed

The warning signs proved so severe that Altman downplayed them while justifying their occurrence. When senior researchers departed the company because of its commercial transformation, Altman explained their departure as regular career evolution rather than a disagreement about the company's direction.

The Critics' Concern:

The safety researchers, together with ethicists at Microsoft, expressed concern about the rapid deployment of OpenAI technology through Bing search and other products. The team faced a challenge because Microsoft deployed the technology at a faster rate than the researchers felt they could perform safety tests to prevent harm and misinformation spread.

Altman's Response:

Altman concentrated on highlighting the positive aspects of the partnership by showing user enthusiasm for Bing features and positive reviews of the technology instead of addressing the safety concerns. Then, he used positive user feedback to prove that critics were mistaken about their safety worries.

The Problem:

Altman demonstrated confirmation bias through his search for *evidence that supported his desired outcome* (the successful partnership) while ignoring valid safety concerns. That users enjoyed the new features did not solve the fundamental safety problems.

A Clearer Example: The CEO of a car company would respond to safety concerns about a new model by focusing on customer design and performance appreciation instead of fixing the safety issues in accidents.

The internal company culture started prioritizing corporate demands over safety-oriented research approaches. The company pressured scientists who advocated for slower deployment and more extensive safety testing by insisting that competition meant they needed faster timelines.

The Reckoning

OpenAI received growing condemnation in 2024 because it released AI products prematurely while neglecting safety standards, which prioritized commercial achievements over its original promise of responsible development. The combination of corporate partnership with research safety proved too difficult for Altman to balance effectively, which resulted in fundamental breaches of the organization's original values.

This scenario illustrates how mission-driven leaders use confirmation bias to justify choices that violate their declared values.

A Note on NASA's Issues

NASA lost its critical thinking foundation during the political and bureaucratic changes that occurred throughout its history. The combination of decreasing public interest and budget limitations forced

NASA to develop a decision-making process that depended heavily on contractors and internal consensus instead of engineering rigor, which resulted in procedural inertia and reduced empirical questioning.

Throughout its history, NASA experienced repeated instances of groupthink and closed mindsets, which stopped engineers from questioning flawed assumptions or expressing safety concerns. The Challenger and Columbia disasters happened because NASA failed to allow internal dissent and underestimated risks and suppressed engineer concerns. Investigations revealed multiple safety improvements remained unimplemented because NASA had failed to learn from past disasters.[OBJ]

NASA's increasing reliance on SpaceX and Elon Musk for launches and infrastructure development has reduced its ability to develop independent technical solutions and maintain oversight, which has compromised NASA's scientific judgment by prioritizing commercial speed and visual appeal. NASA's ability to critically evaluate projects and maintain safety standards deteriorated because of both budget cuts and workforce reductions and the dismantling of oversight bodies, which created an *organizational silence that discouraged dissent.*

The most critical need for assumption challenges was absent from NASA because the organization moved toward political convenience and outside control, which abandoned its traditional skeptical approach and safety-oriented decision process.

Bernie Madoff's Ponzi Scheme

The financial scam perpetrated by Bernie Madoff turned out to be the largest in history because he stole $65 billion from thousands of investors over twenty years. How people missed the deception is interesting to detangle. The extent of the theft stands out, but people missed many warning signs before discovering the full scale of the deception.

The Man Who Tried to Warn Everyone

Harry Markopolos performed a quick investigation that exposed potential issues in Madoff's investment return structure. His thorough four-hour examination established with certainty that the operation was a scam. In 2000, 2001, and 2005, he submitted detailed reports to the government with thirty specific "red flags" that proved Madoff's returns were impossible; nobody listened.

Red Flags That Should Have Stopped Everyone

Madoff's investment strategy demonstrated unrealistic consistency, as he reported financial losses in only seven months over a span of fourteen years, and those losses were so small that they could be deemed negligible. A steady upward trajectory of his returns defied natural investing patterns because such straight-line growth never occurs in actual investment markets.

When questioned, Madoff responded coldly to potential investors who asked too many questions about his investment methods. His response was "take it or leave it," since he believed his strategy would be copied by others after disclosure.

They Trusted Reputation Over Facts

One unfortunate mistake is that size and reputation took over critical thinking and led to this horrific loss of not only money but also life. Suicides followed the Madoff debacle, and lawsuits brought incredible information to light. A single investor said, "*Doubt Bernie Madoff? Doubt Bernie? No*, you *doubt God. Your doubt of God is acceptable, but your doubt of Bernie is unacceptable.*"

How could one man be considered godlike? Friends were recommending the fund to other friends, and the network was being spread as though it were a special secret given only to a select few. Cleverly,

Madoff generally refused to take people's money to invest, and they had to plead with him to do so. This ploy, undoubtedly, led to its increasing charisma.

What was the solution that kept some people solvent while others lost everything? The survivors among the victims kept their financial savings separate from the Madoff investment because they only put a portion of their assets with him.

How to Protect Yourself in the Future

Investors should always demand full transparency about their investments. Steer clear of working with any advisor who cannot provide straightforward explanations about their investment approach. Every investment can and will be affected by market and economic fluctuations.

Be Suspicious of Perfect Returns

Guarantees of unrealistically high returns are a clear warning sign. The prolonged delivery of consistent returns as claimed by Madoff is just as improbable as his initial unrealistic returns. And the auditor also has to be someone from a legitimate company. An investment company handling billions would employ a reputable, nationally recognized auditing firm. A tiny or unknown auditor should raise significant concerns about the investment company. Madoff used a tiny company.

The Bernie Madoff scandal demonstrates that *critical thinking is a fundamental survival tool* for anyone managing investments. The victims who lost everything were intelligent individuals *who failed to ask appropriate questions* at the most critical moment.

The Common Patterns

Across these cases, several common patterns of confirmation bias emerge:

Selective Attention to Evidence: All of these leaders/corporations consistently paid more attention to information that supported their beliefs while giving less weight to contradictory information. This wasn't necessarily conscious deception—they genuinely seemed to deem supporting evidence more compelling and memorable than contradictory evidence.

Reinterpretation of Negative Information: When confronted with information that clearly contradicted their beliefs, these leaders often found ways to reinterpret that information as either irrelevant or actually supportive of their position. Technical problems became temporary obstacles, financial losses became necessary investments, corruption allegations became political attacks, and policy failures became implementation problems.

Attribution of Criticism to Bias: All the leaders showed a tendency to dismiss criticism by attributing it to the biases or vested interests of their critics rather than seriously considering the substance of the criticism. This allowed them to maintain their beliefs without engaging with potentially disconfirming evidence.

Echo Chamber Creation: Confirmation bias led them to surround themselves with people who shared their beliefs and to marginalize or remove those who challenged them. This created an environment where **confirmation bias could flourish unchecked.**

Escalation of Commitment: As evidence mounted against their core beliefs, all of them doubled down rather than reconsidering their positions. This escalation of commitment is a **common consequence** of confirmation bias—the more we invest in a belief, the harder it becomes to abandon it *even when evidence suggests we should.*

The Psychological Roots

Understanding why confirmation bias is so powerful requires recognizing its psychological functions. Our beliefs aren't just abstract ideas—they're often central to our identity, our relationships, and our sense of purpose. Challenging deeply held beliefs can feel like *challenging who we are as people*.

For leaders like those in our examples, their core beliefs were intimately connected to their professional identity and mission. Holmes saw herself as a revolutionary entrepreneur who would transform healthcare. Neumann viewed himself as a visionary leader creating the future of work. Blatter believed he was democratizing world soccer.

Admitting that these beliefs might be wrong would have required not just intellectual adjustment but fundamental identity reconstruction. The psychological pain of this kind of change can be so severe that *our minds automatically protect us from it* by filtering out threatening information.

Breaking Free from Confirmation Bias

Recognizing confirmation bias in ourselves requires deliberate effort and systematic approaches:

Actively Seek Disconfirming Evidence: Instead of looking for information that supports our beliefs, we should actively search for *evidence that might contradict them*. This feels uncomfortable, but it's essential for critical thinking.

Consider Alternative Explanations: When something seems to confirm our beliefs, we should ask ourselves what other explanations might account for the same evidence.

Cultivate Intellectual Humility: Recognizing that we might be wrong about important issues isn't a sign of weakness—it's a sign of intellectual maturity and the foundation of good critical thinking.

Create Diverse Advisory Networks: Surround yourself with people who will challenge your thinking rather than just affirm your existing beliefs.

Use Structured Decision-Making Processes: Formal processes that require considering multiple perspectives and evaluating evidence systematically can help counteract our natural biases.

The leaders in our examples all started with admirable goals and genuine talents. Their failures weren't due to lack of intelligence or malicious intent, but rather to the very human tendency to see what we want to see and ignore what we prefer not to acknowledge. By understanding how confirmation bias operated in their thinking, we can better recognize and combat it in our own decision-making.

Sources and References

To verify any information, we would need to independently research and verify these details using primary sources. Here are the types of sources you would want to consult for each case:

For Elizabeth Holmes and Theranos:

John Carreyrou's reporting in The Wall Street Journal (2015-2018)

Court documents from United States v. Elizabeth Holmes

SEC enforcement actions against Theranos

Carreyrou's book "Bad Blood: Secrets and Lies in a Silicon Valley Startup"

For Adam Neumann and WeWork:

WeWork's S-1 filing with the SEC (2019)

Financial reporting from The Wall Street Journal, Bloomberg, and Financial Times

"The Cult of We" by Eliot Brown and Maureen Farrell

Business news coverage from 2019-2021

For Sepp Blatter and FIFA:

FBI and Swiss law enforcement documents

Court filings in FIFA corruption cases

Reporting from The Guardian, BBC, and other international news outlets

FIFA's own governance documents and reform reports

Chapter 3: The Information Overload Challenge

Rachel's regular schedule began with coffee and reading the newspaper before beginning her workday. She receives 47 phone alerts during her morning as Instagram scrolls while she looks at three news applications during breakfast and two podcasts play during her commute before consuming more daily information than her grandparents did in a week before 9 AM. The information overload makes her feel just as overwhelmed as I do, which explains why she has issues with keeping her focus.

After lunch, Rachel develops mental exhaustion from dealing with opposing retirement fund advice and screen time restrictions for her child. She is drowning in information as she desperately searches for authentic solutions to her valid questions.

People today encounter these primary challenges of modern times because they have unlimited information, yet they lack the wisdom to handle it. Mastering the skill of choosing valuable information for your mental energy stands as the solution over data consumption. The goal is selectivity, so we will explore this approach in this chapter.

The Information Explosion Reality Check

Let's put the numbers into perspective. People in 1986 had access to the equivalent of 40 newspapers of daily information. The amount of daily information you experience today equals 174 newspapers before TikTok and ChatGPT appeared. You would have unlimited access to all newspapers combined with TikTok content along with any Internet AI search results. But the human brain has remained unchanged since 1986 which makes it impossible to ignore this basic truth. We now face an impossible situation. We must *focus on information selection* to determine what needs our attention versus what we can disregard as irrelevant noise.

What this looks like in real life:

The number of shows you want to watch on Netflix has reached 47 while your Watchlist continues to expand.

Out of the 312 people you follow on Instagram you only genuinely care about 20 followers. Why so many "follows" then?

You subscribe to 8 newsletters, but you rarely check any of them. Are you paying for any of them?

Your computer displays 23 active browser tabs in front of you (we both know this is true). Memory overload here?

You have bookmarked 156 articles in your browser, which you planned to read in the future. When will you have time?

Your daily phone notifications amount to 80 on average. Your phone receives excessive calls throughout the day unless you are receiving an emergency call. You receive this number of notifications during your typical day rather than on any specific day. You may get more on any one day.

You need to use your knowledge to make smart choices. Your professional success and wellness, together with financial stability, personal connections and wireless provider choice, all depend on these vital factors. Stress arises because people feel constantly out of touch with important events. As a non-circus performer, you cannot manage multiple balls simultaneously in an effective way. Something has to give.

The Quality vs. Quantity Problem

Too much information fails to improve our decision-making skills directly. When information reaches a certain threshold, it leads to both decreased decision confidence and worse decision accuracy.

The Jam Study Effect in Your Life

Sheena Iyengar demonstrated through research how customers faced with 24 jam options made fewer purchases compared to those who chose from 6 jam varieties. The paradox of choice creates its effect in every setting throughout all environments.

The six-month career transition research by Jake required him to read multiple specific books along with various online tests. With this massive amount of new information he learned it only made his confusion worse. Three specific career transition options helped Lisa achieve her career change within two weeks, even though Jake continued researching without success.

During her Google search for "best diet," Maria found 847,000 results showing keto diets and vegan diets and intermittent fasting diets. The articles presented scientific evidence in their claims. The multiple diet choices she faced caused her to freeze up, which prevented her from changing her poor eating habits during the twelve months she spent on research. When was it time to stop, and how would she know?

The 401k plan of Tom's workplace offered 47 different investment choices for its members. Research proves that people dealing with similar decisions end up picking inferior options or making no choice at all. His coworker achieved better investment performance from his basic investment choices because his plan had only ten carefully designed investment options. *Excessive selection options may lead to diminished success.*

The Source Credibility Crisis

The news sources available to your grandmother during her time consisted of the local paper along with three TV channels and occasional magazine subscriptions (with TV channels turning off at 9 PM). Professional editors, the sources revealed, performed fact-checking and kept their honest reputations. The sources may have held opposing viewpoints, yet their content showed transparency about their information delivery.

People who own smartphones today possess the capability to create authoritative content. Video production becomes simple through voice cloning and text-to-voice technology, which allows anyone to create fake voice statements. The doctor, who dedicated twenty years to his medical practice, now uses the same platform as the teenager who demonstrates good video skills. Who do you trust?

Real-world source confusion examples

The Wellness Influencer versus the Physician:

Mariah struggled with fatigue. The doctor ordered blood tests but noted that stress could potentially be the root cause. With 500K followers, the wellness influencer diagnosed "adrenal fatigue" and then recommended $200 worth of supplements. The influencer maintained a regular posting schedule through powerful story-telling combined with many testimonials from viewers. During his rushed speech, he used complex medical terminology. Mariah spent six months following the influencer's advice before doctors diagnosed her with thyroid problems, which required proper medical intervention. Was the time she spent worth it, or did she endanger her health by following that influencer? The answer seems clear here.

The Financial Guru versus the Financial Planner:

The YouTube financial guru who promoted aggressive day trading strategies caught Mike's attention. During his presentation, the guru showed pictures of his successful deals and luxurious vehicles. The certified financial planner suggested that customers should invest in index funds through a methodical and conservative approach. During the one-year period, Mike lost $15,000 after following the YouTube guru's advice. During this period, the planner's clients generated an average return of 8 percent.

The Political Pundit versus the Policy Expert:

Lisa needed to understand the school funding ballot initiative that appeared on her local election ballot. A political commentator used a three-minute video to explain the ballot measure by making straightforward "good" or "bad" evaluations. A policy researcher developed an exhaustive evaluation that analyzed multiple perspectives. The commentator obtained 50,000 viewer counts. The researcher reached 200 people. Lisa believed the commentator's video but failed to review the researcher's analytical findings. Numbers serve as a poor indicator of anything except showmanship because they cannot demonstrate real value.

Acquiring skills to identify trustworthy sources does not need a research background.

The SIFT Method (Stop, Investigate, Find, Trace, Mike Caulfield):

A pause should come before you start reading or believing or sharing information. The question should be whether you require this information. Does the information help me make better decisions through knowledge-based choices?

Before moving on to the article, you should verify if the coffee health effects information will influence your decisions. Research presents two conflicting perspectives regarding how coffee affects your health. Does anyone check research findings before continuing to read the article? They keep it for future investigation, yet they lack the ability to do so immediately. I drink two cups of coffee per day while maintaining good health. Does my health status need additional information about this?

Spend 30 seconds figuring out who created this information in the source investigation.

Good signs:

- Author has relevant credentials or experience
- Publication has editorial standards and corrections policies
- Claims are backed by specific evidence, not just opinions
- Source acknowledges uncertainty and limitations

Red flags:

- Anonymous authors or vague credentials ("health expert")

- Sensational headlines designed to provoke strong emotions (clickbait)

- Claims that seem too good/bad to be true

- The source avoids presenting uncertainties, complex viewpoints, and any form of doubt.

- Multiple sources should be explored when you need to understand different viewpoints about a topic that matters to you.

When you study health topics, you should look at these types of sources:

- Peer-reviewed research (even if you just read the summary): Software can provide you with a summary of any article.

- Patient information from medical institutions

- Reputable health journalists who specialize in the topic

- The statements made by specialists who work as physicians in their field

- The source tracking technique should be used to verify statements that are beyond ordinary beliefs.

The article headline makes a false statement about wine having superior advantages than physical exercise yet there is no supporting study. You track down the original research through the source tracking method, which revealed that the study analyzed wine-drinking subjects against people who spent most of their time sitting instead of exercising. The misleading headline once again proved to be clickbait content.

Building Your Personal Information Diet

Your health depends on having a balanced information diet, just like you need a balanced food diet. The selection of information prior to consumption forms a fundamental element of this strategy.

The Information Triage System

Level 1: Essential Information

This information serves two purposes: decision-making and safety protection. This includes:

The following information applies to your community needs and workplace responsibilities.

- Health information related to conditions you actually have
- Financial information relevant to your specific situation
- Skills training for your career or serious hobbies

Level 2: Useful Context

The information at this level provides you with context about your world, though you do not need to take immediate action.

- Background on major news events

- Industry trends in your field

- Cultural and scientific developments that interest you

Level 3: Entertainment

Using information serves to entertain and relax as well as to facilitate social connections:

- Sports updates

- Celebrity news

- Funny videos

- Social media updates from friends

- Your preferred sports team's news coverage functions as a suitable form of entertainment.

- A daily news habit, including 12 sports teams and fantasy leagues and sports commentaries, exceeds the recommended amount of information.

Level 4: Junk Information

The information lacks useful content while generating negative emotions.

The combination of political content that makes you angry yet impossible to act upon and disaster reports about locations you will never visit

- Outrage-inducing political content you can't act on

- Disaster news from places you'll never visit

- The news about celebrities is known as celebrity drama

- Social media shows random people arguing

Creating Your Information Rules

You will dedicate 15 minutes to news consumption followed by 10 minutes on social media before you need to stop.

I will reduce the number of news sources I follow to three instead of keeping fifteen.

My goal is to learn about investing, but *I will not attempt to master all three subjects at once*. The human mind cannot achieve expertise in every area simultaneously. You have no requirement to attempt mastery of all these subjects. Your purpose in such a case would remain unclear. The question becomes particularly intriguing to ponder regarding multiple subjects.

I will eliminate all information sources that *keep triggering my anger and anxiety* because these sources do not provide functional solutions. The constant process of getting upset daily proves detrimental to health, so select the beneficial path, which is to minimize these situations as much as possible.

Practical Exercises: Taking Control of Your Information Flow
Exercise 1: The Information Audit

Monitor everything you read or watch throughout three consecutive days.

Record all your information sources.

Note down how much time you spent on these activities.

Record all your feelings during this period.

Take note of the steps you undertake because of the information received.

Real example results:

The ninety minutes spent reading about an unchangeable political scandal kept me angry all day without any productive activity. I spent 5 minutes researching a local candidate I can actually vote for.

Exercise 2: The Source Credibility Check

Choose five regular sources that deliver your information. Each source requires thorough investigation.

Research each of the following items:

The funding source behind this publication remains unknown.

What makes the source qualified to discuss their topics?

The way they perform corrections needs evaluation.

The relevant field experts show respect for this organization.

Real-world discoveries:

The health blog functions as a business that promotes supplements through its author.

The financial newsletter provides reliable forecast predictions because of its established track record.

Your trusted news source distributes content that lacks evidence from other media platforms.

Exercise 3: The 24-Hour Rule

You should hold off from sharing content on social media or discussing articles with others for twenty-four hours during this week. Ask yourself:

The value of this information remains the same to you.

I have confirmed the information to be accurate.

This action will either bring helpful value or simply produce additional noise.

When people use this rule, they reduce their information sharing by approximately 70% according to normal results.

Dealing with Information FOMO (Fear of Missing Out)

People avoid better information habits because they fear missing essential details. And people perceive the fear of missing important information as genuine, although it does not have any strong logical basis.

Reality check questions:

News stories have not brought you any harm for how long?

The frequency of important events does not reach you through different communication channels.

What would be the worst possible result from checking news only once daily instead of monitoring it continuously?

FOMO reduction strategies:

Multiple people in your life will share genuine, important information with you.

1. Your duty to monitor everything remains unnecessary.

2. Organize your information consumption into dedicated blocks of time for news, social media and research instead of continuous day-long browsing.

3. The test is, "What can I do with this information?" you ask yourself. The information becomes unnecessary to you if your answer is "nothing."

The Notification Revolution You Need

Your phone generates *constant interruptions*. Everyone needs to control his or her notifications to manage excessive information.

Notification categories

Keep these on

- Calls and texts from important people

- Calendar reminders for actual appointments

- Banking alerts for unusual activity

- Weather warnings for your location

Turn these off

- Social media likes and comments.

- Everyone should disregard normal news alerts unless there is a location-specific emergency.

- These apps try to persuade customers to buy products repeatedly through their shopping applications.

- People should check their email manually instead of using email notifications.

Real transformation story:

David reduced his stress levels after he eliminated all his non-essential notifications. The reduction of his daily information checks to 67 times allowed him to limit his information source checks to scheduled hours. He then developed personal rules to manage information intake after achieving better work performance and spending more quality time with his family.

Building Your Information Processing Skills

The 5-Minute Rule:

Devote five minutes to orientation before moving forward with a deep information examination.

Take a closer look at the source you are using at present.

What level of expertise does the author have along with their potential biases?

The author presents specific points to you that they want to convey in their message.

You need to create particular questions while reading to enhance your comprehension of the material.

The Summary Test:

You need to restate the fundamental information from lengthy content in your own words. If you cannot explain the information, you should neither take any action on it nor share it with others, as this suggests that you do not fully understand the material. Assess how multiple sources would enhance your comprehension of particular subjects. The information in one source may not be enough; thus, you should examine two or three sources. I always follow the practice of reviewing at least three sources when I find a particular subject difficult to grasp. And sometimes I use more source material.

The "So What?" Question:

When assessing any information, you need to ask, "So what? Does this information alter your thoughts or actions? You should evaluate whether the information merits your time when you fail to answer this question.

Skills Checkpoint

Make sure you grasp all fundamental concepts before proceeding.

Using the SIFT (**S**top, **I**nvestigate the source, **F**ind better coverage, and **T**race claims, quotes, and media) method allows you to evaluate the credibility of information sources within one minute.

__Manage information consumption: Create personal rules for *what information you consume and when.*

__The practice of information triage allows you to organize information into vital and useful sections as well as trivial and unnecessary sections.

__ Control information flow: Have adjusted notifications and sources to reduce being overwhelmed.

Self-Check Questions:

Which three information sources do you use most often?

1. What steps did you use to validate their credibility?

2. Describe your current information diet. What changes would improve it?

3. When was the last time consuming information helped you make a better real-world decision?

Real-World Application Challenge:

Create a major transformation to your information consumption process during the upcoming week.

Unsubscribe from sources that consistently waste your time.

Set specific times for checking news and social media

The main goal is to disable alerts that do not matter.

The SIFT method should be applied to three information sources that you use on a regular basis.

Monitor how these adjustments affect your stress levels and your productivity and decision-making quality. Be surprised at the outcome.

The goal is to prevent information isolation by practicing deliberate information intake instead of spontaneous reactions. You gain control over information overload when you **begin your day with purpose** instead of random data streams.

Your current information habits create an environment for developing skills in distraction-breaking question-asking that reveals vital truths about your personal life.

Chapter 4: Evidence Detective Work

Dr. Martinez had been practicing medicine for fifteen years when a patient named Janet walked into her office clutching a stack of printouts. *"Doctor, I've done my research,"* Janet announced, spreading the papers across the desk. *"These studies prove that my symptoms are caused by chronic Lyme disease, and I need long-term antibiotics."*

Now Dr. Martinez looked at the papers: a blog post about Lyme disease, a testimonial from someone who felt better after antibiotic treatment, a study that had been retracted for falsified data, and an article from a website selling Lyme disease treatments. Janet had spent hours "researching," but she hadn't learned to distinguish between good evidence and convincing-sounding claims.

This scene plays out millions of times daily. People make important decisions about their health, finances, careers, and relationships based on what they think is solid evidence, when they're actually relying on anecdotes, outdated information, or outright misinformation.

Learning to evaluate evidence isn't just for scientists and doctors—it's an essential life skill for anyone who wants to make better decisions in a world full of competing claims.

Understanding the Evidence Landscape

Not all evidence is created equal. Some evidence is like a sturdy bridge you can confidently walk across, while other evidence is like a rope bridge in a windstorm—it might hold, but you wouldn't want to bet your life on it.

The Evidence Pyramid: From Weakest to Strongest

Level 1: Personal Stories and Testimonials (Weakest)

These are individual accounts of what happened to one person.

Example: *"My uncle smoked two packs a day and lived to 95, so smoking can't be that dangerous."*

Why it's weak: One person's experience doesn't tell us what's likely to happen to others. Your uncle might have had exceptional genes, unusual lifestyle factors, or just been very lucky.

When testimonials are useful: They can help you understand what an experience feels like or identify questions to investigate further, but they shouldn't drive major decisions.

Level 2: Expert Opinion (Stronger)

These are conclusions drawn by people with relevant expertise and experience.

Example: A financial advisor with 20 years of experience says, *"Based on my experience, most people should have six months of expenses in emergency savings."*

Why it's stronger: Experts have seen many cases and can recognize patterns, but their opinions can still be influenced by their particular experiences or biases.

When expert opinion is useful: When you need guidance on complex topics and when multiple experts agree.

Level 3: Observational Studies (Stronger Still)

These studies observe large groups of people over time to identify patterns.

Example: A study following 50,000 people for 10 years finds that those who eat more vegetables have lower rates of heart disease.

Why it's stronger: Large numbers reduce the chance that results are due to coincidence, but these studies can't prove that vegetables caused the lower heart disease rates—maybe people who eat vegetables also exercise more or have better healthcare.

When observational studies are useful: For understanding associations and patterns, especially when controlled experiments aren't practical or ethical.

Level 4: Controlled Experiments (Strongest)

These studies test specific interventions by comparing groups that are as similar as possible except for one factor.

Example: A study randomly assigns 1,000 people to either take a new medication or a placebo (fake pill) for six months, then compares outcomes.

Why it's strongest: By controlling other factors, these studies can show whether the intervention actually caused the observed effects.

When controlled experiments are most valuable: For testing whether specific treatments, programs, or interventions actually work.

Real-World Evidence Evaluation in Daily Life

Health Claims: Separating Hope from Hype

Scenario: You see an article titled "Miracle Supplement Reduces Joint Pain by 80%!"

Evidence detective questions:

- What type of evidence is this based on? (Testimonials, small study, large controlled trial?)

- Who funded the research? (The supplement company, independent researchers, government agency?)

- How many people were studied, and for how long?

- What exactly was measured? ("Feeling better" is subjective; specific mobility tests are more objective)

- Were there control groups? (Did some people get fake supplements to compare?)

- Have other researchers found similar results?

Red flags:
- Study funded by the company selling the product

- Based on testimonials rather than controlled research

- Claims of "miraculous" results

- No mention of side effects or limitations

- Results that seem too good to be true

- Better evidence would look like:

- Independent research not funded by supplement companies

- Studies comparing the supplement to placebos

- Research published in peer-reviewed medical journals

- Multiple studies showing similar results

- Honest discussion of limitations and side effects

Investment Advice: Following the Money Trail

Scenario: A financial guru claims, "I've discovered a simple strategy that beats the stock market 90% of the time!"

Evidence detective questions:

- What evidence supports this claim? (Backtesting on historical data, real trading results, academic research?)

- How long was this strategy tested? (Strategies that work for one year might fail over five years)

- What were the conditions during the testing period? (Bull market, bear market, economic stability?)

- Were transaction costs and taxes included in the results?

- How many people have tried this strategy, and what were their results?

- What are the risks and worst-case scenarios?

Red flags:

- Claims of consistently beating the market

- No discussion of risks or losing periods

- Results that conveniently exclude fees and taxes

- Pressure to act quickly before the "opportunity" disappears

- Testimonials from unnamed successful investors

Better evidence would include:

- Long-term performance data including bear markets

- Transparent reporting of all costs and taxes

- Independent verification of results

- Academic research on similar strategies

- Honest discussion of risks and volatility

The Correlation vs. Causation Trap

One of the most common evidence mistakes is assuming that because two things happen together, one must have caused the other. This confusion leads to countless bad decisions.

Real-world examples of correlation without causation:

The Ice Cream and Drowning Connection:

Data shows that ice cream sales and drowning deaths both increase during the same months. Does ice cream cause drowning? Of course not—both increase during summer when people swim more and eat more ice cream.

The Divorce Rate and Margarine Consumption:

For several years, the divorce rate in Maine correlated almost perfectly with per capita margarine consumption. This is pure coincidence—neither causes the other.

The College Education and Income Link:

People with college degrees earn more money on average than those without degrees. But does the degree itself cause higher earnings, or do people who go to college have other advantages (family support, networking, personal motivation) that would lead to higher earnings anyway?

How to think about causation more carefully:

Ask these questions:

- Could something else explain both observations?

- Is there a plausible mechanism for how A could cause B?

- Does the timing make sense? (Causes should come before effects)

- Have researchers tried to control for other factors?

- Are there examples where A happens without B, or B happens without A?

Example: Exercise and Mood

Observation: People who exercise regularly report better moods.

Possible explanations:

- Exercise causes better mood (through brain chemistry changes)

- Better mood causes more exercise (people feel motivated to be active)

- Something else causes both (good health, financial security, social support)

- The relationship works both ways

Better evidence would include: Studies that randomly assign some people to exercise programs while others don't, controlling for other factors, and measuring mood changes over time.

Statistical Literacy for Real People

You don't need to become a statistician, but understanding a few basic concepts will help you evaluate numerical evidence more wisely.

Understanding Percentages and Baselines

Misleading claim: "This treatment reduces your risk of heart attack by 50%!"

The detective work: 50% of what? If your baseline risk was 2 in 1,000, a 50% reduction means your risk drops to 1 in 1,000—a much smaller absolute change than it sounds.

Better way to present this: "For every 1,000 people who take this treatment, one fewer person will have a heart attack compared to those who don't take it."

Sample Size Matters

Misleading claim: "100% of people in our study lost weight on this diet!"

The detective work: How many people were in the study? If it was only 3 people, 100% means all 3 people—not very convincing evidence.

Rule of thumb: Be more skeptical of dramatic results from small studies. Large, consistent effects might be meaningful even with smaller samples, but small effects need large samples to be believable.

The Average Can Be Deceiving

Example: A company reports that its employees' "average" salary is $85,000.

The detective work: This could mean:

- Most employees earn around $85,000 (if salaries are evenly distributed)

- A few executives earn $500,000 while most employees earn $40,000 (if the CEO's salary pulls up the average)

- Better questions: What's the median salary? (The amount

earned by the person in the middle) What percentage of employees earn above and below $85,000?

Practical Evidence Evaluation Tools
The SIFT Method for Quick Assessment (Caulfield)

Stop: Before accepting or sharing information, pause.

Investigate the source: Who created this information? What are their credentials and potential biases?

Find better coverage: Look for multiple sources and different perspectives on the same claim.

Trace claims to their origin: Follow extraordinary claims back to their original source.

The Evidence Quality Checklist

When evaluating any claim, ask:

- Source credibility: Who is making this claim? What expertise do they have? What might motivate them?

- Evidence type: Is this based on personal stories, expert opinion, observational studies, or controlled experiments?

- Sample size: How many people or cases does this evidence include?

- Replication: Have other researchers found similar results?

- Plausibility: Does this claim fit with what we know about how the world works?

- Transparency: Are the methods and data clearly described?

- Conflicts of interest: Who funded this research? Who benefits if people believe these claims?

Evaluating Product Reviews

Scenario: You're buying a laptop online and reading customer reviews.

Evidence detective approach:

- How many reviews are there? (10 reviews vs. 1,000 reviews)

- When were the reviews written? (Recent reviews might reflect current product quality)

- What specifically do reviewers mention? (Detailed experiences vs. vague praise/complaints)

- Are there patterns in complaints? (Multiple people mentioning the same problems)

- Do any reviews seem fake? (Overly enthusiastic language, generic comments, all posted on the same day)

Evaluating News Claims

Scenario: You see a headline: "New Study Shows Coffee Prevents Cancer!"

Evidence detective approach:

- Read beyond the headline to the actual study details

- Check if this was an observational study or controlled experiment

- Look at the sample size and how long the study lasted

- See if other news sources are reporting the same findings

- Check if medical experts are commenting on the study's limitations

- Consider whether this one study contradicts or confirms previous research

Evaluating Career Advice

Scenario: A career coach claims, *"Following your passion always leads to career success!"* Not necessarily so.

Evidence detective approach:

- What evidence supports this claim? (Success stories, systematic research, expert analysis?)

- How is "success" defined? (Happiness, income, career advancement, work-life balance?)

- Are there counter-examples? (People who followed their passion and struggled financially?)

- What other factors might influence career success? (Skills, market demand, networking, timing)

- Does this advice consider different personality types and life circumstances?

When Evidence Conflicts: Making Decisions with Uncertainty

Sometimes you'll find good evidence pointing in different directions. This is frustrating but normal—many important questions don't have clear-cut answers.

Strategies for dealing with conflicting evidence:

Look for areas of agreement: Even when experts disagree on details, they often agree on broader principles.

Example: Nutrition experts disagree about specific diets but generally agree that eating more vegetables and less processed food is beneficial.

Consider the consequences of being wrong: If the downside of following advice is small, you might try it even with uncertain evidence. If the downside is large, wait for stronger evidence.

Example: Taking a vitamin supplement with uncertain benefits but minimal risks might be reasonable. Undergoing surgery based on uncertain evidence requires much higher confidence.

Update your beliefs gradually: As new evidence emerges, adjust your confidence rather than completely flipping your position.

Example: If you initially thought a treatment was 70% likely to work, and a new study suggests it's less effective, you might lower your confidence to 50% rather than concluding it definitely doesn't work.

Practical Exercises for Evidence Evaluation

Exercise 1: The Daily Evidence Audit

For one week, choose one piece of information you encounter each day (news article, social media post, advertisement, advice from a friend) and evaluate it using the Evidence Quality Checklist. Write down:

- What claim is being made?

- What type of evidence supports it?

- What questions would you need answered to be more confident?

- How credible is this evidence?

Exercise 2: The Health Claim Investigation

Choose three health-related claims you've heard recently (about supplements, diets, exercise, or medical treatments). For each claim:

- Find the original research behind the claim

- Identify what type of study it was

- Look for other studies on the same topic

- Read what medical experts say about the evidence

- Decide how confident you are in the claim

Exercise 3: The Correlation Hunt

For one week, notice when you or others assume causation from correlation. Examples might include:

- *"Students at private schools get better test scores, so private schools are better"*

- *"People who drink wine live longer, so wine is healthy"*

- *"Successful people wake up early, so waking up early causes success"*

For each example, brainstorm alternative explanations for the correlation.

Chapter 4 Skills Checkpoint

Before moving on, ensure you can:

__Distinguish evidence types: Recognize the difference between testimonials, expert opinions, observational studies, and controlled experiments

__Evaluate source credibility: Quickly assess who is making claims and what their motivations might be

__Understand correlation vs. causation: Recognize when two things happening together doesn't mean one caused the other

__Apply basic statistical thinking: Ask good questions about sample sizes, averages, and percentages

__Use practical evaluation tools: Apply the SIFT method and Evidence Quality Checklist to real-world claims

Self-Check Questions:

Describe a time when you made a decision based on weak evidence. How might you evaluate that evidence differently now?

What's an example of correlation that people often mistake for causation in your field or area of interest?

When evaluating competing claims about an important decision in your life, what evidence would you most want to see?

Real-World Application Challenge:

For the next two weeks, before making any significant purchase, health decision, or career move, apply the Evidence Quality Checklist to the information you're basing your decision on. Notice how this changes your confidence in different sources of advice and information.

Remember: the goal isn't to become so skeptical that you never trust any information. The goal is to calibrate your confidence appropriately—being more confident in stronger evidence and more cautious with weaker evidence. Like Dr. Martinez helping Janet learn to distinguish between solid medical research and health misinformation, you can develop the skills to navigate an evidence-rich world more wisely.

With your evidence evaluation skills sharpened, you're ready for the next challenge: learning to spot the logical fallacies that can make weak arguments sound convincing and strong arguments seem doubtful.

Chapter 5: Evidence Detective Work

After fifteen years of medical practice, Dr. Jonas received Janet into her office while Janet brought multiple printouts in her hand. *"These studies prove that my symptoms are caused by chronic Lyme disease, and I need long-term antibiotics."*

Dr. Jonas examined the documents, which included a Lyme disease blog entry together with a patient testimonial, and a retracted study and advertisements for Lyme disease remedies. Janet had devoted numerous hours to research, yet she failed to recognize the difference between authentic evidence and deceptive-sounding statements.

The described situation is repeated thousands of times throughout the day. People use perceived evidence to make crucial choices about their health, finances, careers and relationships despite *using outdated or incorrect information.*

The skill of evidence evaluation is important for everyone because it allows us to make improved choices in situations with many conflicting statements.

Understanding the Evidence Landscape

Evidence levels differ significantly from one another. Some evidence functions as a reliable bridge for walking, yet other evidence functions like a windy rope bridge. The latter might be used in an emergency but would never be the standard one trusted as a daily safe structure.

The Evidence Pyramid: From Weakest to Strongest

Level 1: Personal Stories and Testimonials (Weakest)

The evidence consists of personal narratives about the experiences of one individual.

Consider the statement *"My uncle smoked two packs of cigarettes daily until reaching age 95, so smoking cannot be dangerous"* shows the weakness of his argument.

The weakness of this evidence comes from the fact that *one person's experience fails to predict outcomes* for other individuals. Your uncle lived a long life because of his special genetic makeup combined with other lifestyle elements together with fortunate circumstances.

The usefulness of testimonials lies in helping you experience what something feels like and can be helpful in generating new questions to explore, but they *should not lead to major choices*.

Level 2: Expert Opinion (Stronger)

The conclusions stem from expertise along with the experience of relevant individuals.

A financial advisor, who has spent twenty years in practice states, *"My experience shows that most people need six months of emergency savings."*

The strength of this evidence comes from expert experience, yet their conclusions remain susceptible to their personal experiences and biases. This expert might be biased and doesn't recognize it.

Usefulness of expert opinion emerges in two situations: when complex guidance is required and when multiple specialists reach agreement.

Level 3: Observational Studies (Stronger Still)

These research studies observe extensive groups over time to identify emerging patterns.

A study involving 50,000 participants over 10 years discovered that vegetable consumption is linked to reduced heart disease occurrences. This is a large sample over an extended period and is more likely to produce reliable results than a smaller sample over a limited span of time. What does a large sample size remove?

The study becomes more robust because large participant numbers decrease *the possibility of random coincidence,* yet it cannot establish cause-effect relationships between vegetables and heart disease prevention since participants who eat vegetables might exercise or receive better medical care. So there are other factors to be considered, and it may not be simply vegetables that caused the effect.

Observational studies find their greatest value when researchers need to *identify patterns or associations,* but controlled experiments present both practical and ethical challenges. Remember what I said about correlation? This is where you have to be careful when you begin to see associations because they may be spurious.

Level 4: Controlled Experiments (Strongest)

Research studies that evaluate specific interventions use somewhat identical groups for comparison except for a single variable.

The study selects 1,000 participants for either receiving the new medication or a placebo (fake pill) during six months before measuring results.

These studies maintain the strongest evidence because researchers control other factors to demonstrate how the intervention produces observed effects. The problem remains that although it seems to be providing strong evidence, there still might be other factors for which they failed to account. Usually, the research will indicate this in its discussion of the conclusions.

Real-World Evidence Evaluation in Daily Life

Health Claims: Separating Hope from Hype

A news headline reads, "Miracle Supplement Reduces Joint Pain by 80%!"

The following evidence detective questions were developed to analyze this piece:

The basis for this conclusion *rests on what type of evidence*? (Testimonials, small study, large controlled trial?)

Research funding came from which organization? (The supplement company, independent researchers, government agency?) When you see an indication that a "clinical study was conducted," That's not very reassuring because anyone can underwrite a study, for instance, the supplement company, and say that there was a clinical study conducted. The words "clinical study" mean nothing.

The research involved what number of participants and did they participate for what duration? For research purposes, the number of participants is usually referred to as the "N."

The study measured what specific factors? ("Feeling better" is *subjective*; specific tests with a rating score are *objective*.) For this reason, psychological testing always uses scales that have standardized scores. Subjectivity is a poor measure of anything.

Were there control groups? Some individuals received deceptive supplements for comparison purposes. In fact, not only do we have the placebo effect in experiments, we also have something called the nocebo effect. The nocebo effect occurs when *negative expectations* about treatments or interventions result in actual negative side effects or symptoms even though the *treatment has no known physical effects*. The nocebo effect works the opposite of the placebo effect, meaning that good hopes lead to good outcomes. It's really mind over matter.

Red flags: funding, limitations and reviews

The study received funding from the product-selling company.

The research base consists of testimonials rather than scientific controlled studies.

The treatment promises "miraculous" outcomes.

The report *does not present any details* about side effects together with some limitations.

The results appear to be unbelievable given their nature. I've seen this in a specific weight loss study where a man and his daughter came, and within a week, the man had lost a significant amount of weight, while the daughter lost perhaps 20% of what he lost. They were both on the same diet and engaged in the same activity, but he lost more. Why? It's a matter of biology.

Better evidence would look like:

The supplement industry has no role in funding independent research conducted by different entities.

The studies used the supplement as a treatment while providing some participants with inactive placebos for comparison.

The research was published in medical journals that operate with peer-review systems. Note: Some journals are owned by supplement companies.

Multiple researchers at different institutions, possibly in various areas of the world, conducted studies that showed similar outcomes from this research.

Researchers must acknowledge both the boundaries and adverse effects of their findings.

Investment Advice: Following the Money Trail

The financial expert reveals to investors that his simple trading approach outperforms stock market predictions 90% of the time. You could refer to this as a "quick money" investment. The only thing you may get will be poorer quickly.

Evidence detective questions:

The evidence used to support this assertion consists of what? (Backtesting on historical data, real trading results, academic research?) Where's the data?

The duration of this strategy's testing period was what? A strategy that proves beneficial in one year might fail during a five-year period. That's why we like longitudinal studies better.

The conditions that existed throughout the testing period need to be defined. Was there an infusion of money to some industry by the government, or was there a new IPO?

The researchers should analyze whether economic stability along with bull and bear market conditions existed during the testing period. Here, charts are invaluable.

The research included both transaction costs and taxes when generating its findings. Think in terms of gross profit as well as net profit. In the entertainment industry, they may refer to it as "above the line."

What were the results when multiple people implemented this strategy?

What are the risks and worst-case scenarios?

More red flags:

The study made statements about beating the market consistently.

The document fails to present any information regarding potential risks or negative periods.

The results avoid presenting costs and taxes that might be involved.

The "opportunity" will disappear soon; therefore, you must take immediate action according to the pressure. Have you ever seen an offer that contains a countdown clock on it? The clock is intended to scare you into immediate action.

The investors who were interviewed in the testimonials remain unidentified.

Better evidence would include:

Performance data extending over multiple years with bear market periods included.

The findings show transparent reporting of all taxes and costs.

The results need independent verification to be accepted as valid.

Academic research on similar strategies

The discussion should contain honest information about the risks together with volatility levels.

The Correlation vs. Causation Trap

People tend to mistake the occurrence of two events as proof that one event leads to the other. Such confusion leads to countless bad decisions.

Correlation without causation appears in multiple situations that occur in real-world environments.

The Ice Cream and Drowning Connection:

Yes, here's another ice cream example. The rise in ice cream sales together with the increase in drowning deaths appears during the same seasonal months, according to statistical data. Is drowning caused by ice cream consumption? Of course not—both increase during summer when people swim more and eat more ice cream. That's called a spurious correlation.

Maine recorded an exact correlation between its divorce rates and margarine consumption rates throughout multiple years.

These findings represent a random coincidence since neither factor produces any effect on the other.

The College Education and Income Link:

College-educated people receive higher average earnings than individuals who do not possess degrees. The educational degree by itself fails to explain why people earn more money because students who enroll in college may possess additional advantages, including family backing, professional contacts and internal drive, that would produce higher salaries independently. Also, there may be a corporate bias to hire those with college educations rather than those who don't have them. Anyone who doesn't have a college education, therefore, couldn't make higher earnings because they'd never have the opportunity.

People need to evaluate causation through more thoughtful methods.

Ask these questions:

What other factors could provide alternative explanations for these observations?

Does a plausible mechanism exist to explain how A would lead to B?

Is the sequence of events appropriate? (Causes should come before effects.)

Have researchers attempted to control other influencing elements during their study?

Example: Exercise and Mood

Regular exercise among people produces improved mood states. We've certainly seen a lot of research on this.

Possible explanations:

People who exercise regularly develop better mood states because their brain chemistry undergoes changes.

Better mood leads people to exercise more because they develop the motivation to be physically active.

The study uses alternative variables to investigate how health benefits together with economic stability and social networks affect well-being. Occasionally we have to look at the money factor even in exercise and certainly in mood. Someone working two or three jobs may not have the time or the money to exercise, and their financial concerns will affect their mood whether they exercise.

Something else causes what we're seeing (good health, financial security, social support).

Better evidence would include studies that randomly assign some people to exercise programs while others don't, controlling for other factors, and measuring objective mood changes over time.

Statistical Literacy for Real People

Understanding a few core statistical principles enables anyone to analyze numerical data more effectively, even without becoming a statistician. Don't let statistics scare you. You only need to know a few of them, and they will serve you well.

Understanding Percentages and Baselines

Misleading claim: *"This treatment reduces your risk of heart attack by 50%!"*

The detective work: 50% of what? When your original heart attack probability reached 2 in 1,000, the 50% reduction would reduce your risk to 1 in 1,000, which demonstrates an actual *smaller absolute change.* What's the number we're looking at here?

A more suitable approach to presenting the result would be: *"The study reveals one less heart attack among 1,000 people who use this treatment instead of those who do not."* The problem, of course, is that this is not a solid marketing selling point, and no one's going to use it. The first one seemed phenomenal, if misleading.

Sample Size Matters

Misleading claim: *"100% of people in our study lost weight on this diet!"*

The detective work: How many people were in the study? The study results become invalid when using 100% because the total number of participants may equal only 10 people, thus making the results unconvincing. We see this in political polling all the time. But it's also been used by the diet industry ad nauseam.

You should maintain stronger doubt about impressive outcomes from investigations that involve small participant groups. Large, consistent effects might be meaningful even with smaller samples, but studies and experiments need large samples to be believable. This is especially important when a new medication comes to market. My orientation has always been to wait until the medication has been on the market for several years and thousands of people have used it without great difficulty.

The Average Can Be Deceiving

Example: A company reports that its employees' "average" salary is $85,000.

The detective work could mean:

Most employees earn around $85,000 (if salaries are evenly distributed).

A few executives earn $500,000 while most employees earn $40,000 (if the CEO's salary pulls up the average).

Better questions: What's the median salary? (The amount earned by *the person in the middle*) What percentage of employees earn above and below $85,000?

Practical Evidence Evaluation Tools

The SIFT Method for Quick Assessment

This method (which we outlined in Chapter 3) was developed by Mike Caulfield, a digital literacy expert, and it contains the following:

Stop: Before accepting or sharing information, pause.

Investigate the source: Who created this information? What are their credentials and potential biases?

Find better coverage: Look for multiple sources and different perspectives on the same claim.

The origin of claims should be **traced** to their initial source because extraordinary claims require proof.

The Evidence Quality Checklist

Let's do a quick rundown of some things we **need to always consider** when evaluating any claim, ask:

• Source credibility: Who's making this claim? What expertise do they have? What might motivate them?

• Evidence type: Is this based on personal stories, expert opinion, observational studies, or controlled experiments?

• Sample size: How many people or cases does this evidence include?

• Replication: Have other researchers found similar results?

• Plausibility: Does this claim fit with what we know about how the world works?

• Transparency: Are the methods and data clearly described?

• Conflicts of interest: Who funded this research? Who benefits if people believe these claims?

Evidence Evaluation in Different Life Areas

Evaluating Product Reviews

You may purchase a laptop online while consulting customer reviews during your purchase.

Evidence detective approach:

• How many reviews are there? (10 reviews vs. 1,000 reviews). Also remember, competitors will try to put negative reviews in for any product.

• When were the reviews written? (Recent reviews might reflect current product quality)

• What specifically do reviewers mention? (Detailed experiences vs. vague praise/complaints)

• Are there patterns in complaints? (Multiple people mentioning the same problems)

• Do any reviews seem fake? (Overly enthusiastic language, generic comments, all posted on the same day)

Evaluating News Claims

Scenario: You see a headline: "New Study Shows Coffee Prevents Cancer!" Sounds familiar, right?

Evidence detective approach:

• Read beyond the headline to the actual study details

• Again, check if this was an observational study or controlled experiment

• Look at the sample size and how long the study lasted

• See if other news sources are reporting the same findings. The problem here is that the media picks up press releases and runs them as though they are valid research study findings.

• Check if medical experts are commenting on the study's limitations

• Does this single study establish whether it supports or opposes findings from earlier research?

Evaluating Career Advice

Scenario: A career coach claims, *"Following your passion always leads to career success!"*

Evidence detective approach:

What proof exists to support this claim? (Success stories, systematic research, expert analysis?)

• How is "success" defined? (Happiness, income, career advancement, work-life balance?)

• Are there counter-examples? (People who followed their passion and struggled financially?) I remember speaking to a dentist who said, *"If I could go back and talk to my 18-year-old self, I'd advise against becoming a dentist."*

Other elements besides skills and market demand and networking and timing also determine career success.

This guidance fails to take into account the needs of various personality types as well as life situations. For this reason, many people undergo vocational testing assessments to identify their interests and skills. I recall a friend being told she should be a forest ranger, and when she took another study, they said she should be a car mechanic. Which one should she follow? Not all testing may give you the most helpful results.

When Evidence Conflicts: Making Decisions with Uncertainty

You will encounter good evidence that leads in multiple directions. It is both frustrating and typical that several vital questions lack definitive solutions.

Strategies for dealing with conflicting evidence:

Although experts disagree about particular details, they maintain shared fundamental principles.

The experts disagree about specific diets, but they agree that vegetables and less processed food benefit health.

Weigh the negative outcomes of deciding based on inadequate evidence to determine whether to proceed. When the potential risks are large, it's best to wait until more evidence becomes available.

Taking vitamin supplements with unknown advantages yet minimal adverse effects could be a reasonable decision. The need for surgical intervention, however, requires very strong evidence since the risks are high.

You should make adjustments to your beliefs at a steady pace as new evidence becomes available instead of making complete changes to your position.

After starting with a 70% confidence level in a treatment's effectiveness, a new study showing reduced efficacy would make you decrease your belief to 50% rather than discarding the treatment's potential effectiveness completely.

Practical Exercises for Evidence Evaluation

Exercise 1: The Daily Evidence Audit

Choose one piece of information each day from news articles, social media posts, advertisements, or friend advice for a week to assess using the Evidence Quality Checklist. Write down:

The main assertion presented in this statement stands as follows:

What type of evidence supports it?

You need more information to build stronger confidence about the matter.

How credible is this evidence?

Exercise 2: The Health Claim Investigation

Choose three health-related claims you've heard recently (about supplements, diets, exercise, or medical treatments). For each claim:

Find the original research behind the claim

What type of study it was

Look for other studies on the same topic

The professional medical perspective regarding evidence can be accessed through reading their statements.

Decide how confident you are in the claim

Exercise 3: The Correlation Hunt

Throughout the week observe when people or you establish cause-effect relationships from observed patterns. Examples might include:

Students from private schools achieve better test results, which demonstrates that private schools provide better education.

The fact that people who drink wine tend to live longer proves that wine provides health benefits.

Early risers among successful people lead to their success because early rising causes success.

For each example, *brainstorm alternative explanations* for the correlation.

Chapter 4: Skills Checkpoint

You need to meet the following criteria before continuing:

Distinguish evidence types: Recognize the difference between testimonials, expert opinions, observational studies, and controlled experiments

Evaluate source credibility: Quickly assess who is making claims and what their motivations might be.

Understand correlation vs. causation: Recognize when two things happening together doesn't mean one caused the other.

Apply basic statistical thinking: Ask good questions about sample sizes, averages, and percentages.

Use practical evaluation tools: Apply the SIFT method and Evidence Quality Checklist to real-world claims.

Self-Check Questions:

1. When did you use minimal evidence for your decision? What changes would you make to your evaluation process today?

2. Your field of interest demonstrates how people often incorrectly link cause and effect by using an example of correlation. What specific example would you provide?

3. What evidence would you need to evaluate competing claims regarding an essential life decision?

Real-World Application Challenge:

Apply the Evidence Quality Checklist to every significant purchase, health decision, and career move before you take action during the next two weeks. The application of this process enables you to recognize different levels of trustworthiness in advice and information sources.

Your skepticism should not reach a point where you doubt everything you hear. You should match your confidence levels with evidence strength so you can trust solid evidence while being more careful about less robust evidence. Through Dr. Jonas' guidance, Janet learned to recognize medical research on health misinformation, which you can apply to develop your evidence-based decision-making abilities.

Your evidence evaluation capabilities are now ready to help you identify logical fallacies, which make poor arguments appear strong while solid arguments appear weak.

Chapter 6: Spotting Logical Fallacies in the Wild

A lex considered himself an effective debater. He kept his political discussions sharp while arguing workplace policies and persuaded his friends to dine at new restaurants. But when a question of curfew came up with his daughter, he felt he needed to be firm and that there would be no discussion. No discussion? He never thought that she deserved more credit, and he had to be more flexible. His daughter's observation, *"Dad, other kids' misbehavior doesn't mean I will, too,"* demolished his curfew argument.

It wasn't easy, but Alex understood his daughter was correct, and he had to admit it. The fallacy of hasty generalization occurred in his thinking because he believed particular cases established universal truths. He unknowingly accepted various logical errors in news articles and advertisements and political arguments for multiple years. Alex now questioned everything and needed to reconsider everything.

Alex learned to approach information differently after that conversation took place. He identified several types of arguments that seemed sound but actually rested on logical mistakes. Learning to detect errors proved to him it was not about intellectual showmanship, yet it helped him protect his decisions and prevent deception.

Why Do Smart People Fall for Bad Arguments?

Every day, intelligent, educated people fall victim to this phenomenon. The initial step toward building resistance *requires understanding the reasons behind* the statements.

Fallacies **feel right and work** because they:

• *Tap into emotions*: Fear, anger, hope, and pride can override careful thinking

• *Confirm existing beliefs*: We're more likely to accept arguments that support what we already think

• *Use familiar patterns*: They mimic valid reasoning but with crucial flaws

• *Come from trusted sources*: People accept fallacious reasoning when it comes from respected sources

• *Address genuine concerns*: The underlying worry might be legitimate even when the reasoning is flawed

Intelligent individuals regularly fall victim to logical fallacies, which function as persuasive methods in their daily lives. Understanding their operational mechanism provides the foundation to develop resistance against them.

Real-world example:

A local politician argues, *"My opponent wants to defund the police. Do you want criminals running wild in our streets?"* This argument sounds compelling because it addresses actual concerns about safety, but *it's actually a "straw man" fallacy*—misrepresenting the opponent's position. The opponent might support police reform, not elimination, but *the politician creates a false choice* between current police funding and complete lawlessness.

The Top 10 Fallacies You Encounter Daily

1. **Ad Hominem**: *Attacking the Person* Instead of the Argument

You shift the focus from addressing their argument to *targeting their personal qualities* or personal background or reasons for taking a position.

Real-world examples:

• The vaccine advice from Dr. Smith is worthless because *she appears young and does not have any children.*

• He supports lower taxes *because he is wealthy* and ignores the needs of ordinary citizens.

• She supports remote work because s*he wants to avoid going to the office* because she's lazy.

Why it's fallacious:

1. A **person's background** or personal characteristics do not automatically invalidate their arguments. A person may have dubious intentions, yet their supporting evidence and logical structure remain sound.

When personal factors DO matter:

It becomes acceptable to assess the credibility of a person when their qualifications directly relate to the matter at hand. Doctors require medical training to provide reliable health recommendations. Financial advisors need established success records when people seek investment guidance.

How to respond constructively:

"I want to concentrate on the factual basis of this viewpoint instead of the individual who presented it." What specific points make you skeptical about the original argument?

2. **Straw Man**: Misrepresenting Someone's Position

What it looks like:

The actual statement of the person remains untouched but the opposing side uses a simplified or altered version of their argument for attack purposes. They rework it to fit their needs and their position; in doing so, they distort it. In doing this, they may also try to distract you in another direction and away from the original point that was being made.

Real-world examples:

- Person A: *"I think we should have stricter background checks for gun purchases."*

- Person B: *"So you want to take away everyone's guns and leave us defenseless?"*

- Person A: *"I support universal healthcare."*

Person B: *"So you want the government to control every aspect of our medical decisions?"*

The environmental benefits from eating less meat should be our main focus, according to Person A.

Person B says, *"So you want to force everyone to become vegetarian?"*

Why it works:

The presented version of the argument tends to be more challenging to defend than the actual position because it uses more emotional language.

How to spot it:

The original argument lacks extreme language which appears in the distorted version. Evaluate the original statement to determine *if the presented version reflects its true meaning*.

How to respond:

"I believe your interpretation of my stance is incorrect. What I'm actually suggesting is..."

3. **False Dilemma**: Presenting Only Two Options When More Exist

What it looks like:

The presentation of two exclusive options when multiple alternatives or intermediate choices exist.

Real-world examples:

- The economy exists in a state of either full opening or perpetual lockdowns.

- The choice is simple: you must either join our group or fight against us.

- The lack of oil drilling across the nation would result in zero energy availability.

- The support of military forces stands as the only way to demonstrate patriotism.

Why it works:

The approach makes complex situations easier to understand while making people choose sides which leads them toward the presenter's preferred option.

Real complexity:

Multiple options and middle-ground solutions exist for the most crucial matters in life. The *most effective solutions unite components* from various approaches.

How to respond:

"I believe we should explore alternative solutions beyond those currently presented. Can we explore combinations of different approaches as well as completely new alternatives?"

4. **Appeal to Authority**: Misusing Expert Opinion

What it looks like:

The presentation of truth based on authority statements ignores both the authority's expertise and their credibility level.

Real-world examples:

- A popular actor claims the supplement treated his arthritis so it proves to be effective.

- The dentist recommends this particular investment plan to me.

- Einstein believed in God so we should dismiss atheism as a valid belief.

- My successful friend believes that going to college wastes time for students.

Why it's sometimes fallacious:

Expertise in one field does not automatically grant someone mastery of all domains. Experts who possess relevant knowledge can be incorrect when there is no agreement between experts about a particular subject.

When authority appeals ARE valid:

- The authority possesses appropriate expertise in this matter.

- All experts in that field maintain agreement on this issue.

- The authority bases their assessment on concrete evidence instead of relying on personal anecdotes.

How to evaluate:

Examine whether this person holds expertise in the relevant field. Do other experts agree? What evidence supports their claim?

5. Bandwagon Fallacy: Everyone's Doing It

What it looks like:

Some people use popularity as evidence to support their statements about what is right or good.

Real-world examples:

- Most people back this candidate so you should also support him.

- Cryptocurrency has become popular among all investors yet you remain outside this trend.

- All the trendy businesses have shifted to remote work arrangements.

- Most parents from this school district enroll their children in this tutoring service.

Why it feels convincing:

Human beings have *natural tendencies to conform* to social trends because they are part of social groups. Popular choices appear safer and more valid to people.

When popularity IS relevant:

- You want to maintain social group inclusion.

- Network effects play a crucial role when it comes to social media platforms and communication tools.

- The opinions of locals about restaurants serve as an example of collective wisdom, which can be beneficial.

When popularity ISN'T relevant:

- Questions about truth and facts fall outside the scope of this assessment.

- Moral and ethical choices require separate consideration.

- The crowd does not possess necessary knowledge about this topic.

How to respond:

Popularity does not necessarily indicate something is suitable for our current situation. We need to assess both advantages and disadvantages specifically.

6. Slippery Slope: One Thing Must Lead to Another

What it looks like:

The argument presents an inevitable sequence of negative effects following an initial action without demonstrating actual cause-effect relationships.

Real-world examples:

- Students who get test retake opportunities will begin requesting every grade re-do, leading to a loss of responsibility skills.

- This statement presents two opposing claims: "*Raising the minimum wage will result in business automation that leads to massive job losses*" and "*Allowing kids to play video games for an hour will turn them into permanent gamers who never want to leave the house."*

- The argument relies on a series of consecutive steps that seem inevitable even though numerous intervening factors exist to stop or alter the chain.

The argument becomes valid when:
- The argument becomes valid when there exists proven historical evidence showing similar patterns of progression.

- The chain of events must display logical progression between its individual elements.

- The absence of protective measures allows this type of situation to develop.

How to evaluate:

The evaluation process requires you to verify the probability of each step in this chain. What factors would stop this chain of events from happening? The observed patterns throughout history show similar effects.

7. Red Herring: Changing the Subject

What it looks like:

The introduction of unrelated material serves as a diversion from the core argument or the main inquiry.

Real-world examples:
- The policy change will affect what happens to small business operations.

- The respondent comes from a wealthy background, which makes him unable to understand what working families face.

- The proposed solution lacks direction in addressing customer complaints.

- Since its inception our company has demonstrated dedication to quality by having its founder establish the organization with fundamental values.

- What evidence shows that this treatment will maintain its safety standards?

- Thousands of people need assistance with this condition which drives their desperation for help.

Why it works:

The new topic contains strong emotional or political appeal that allows people to dismiss the original question.

Ask yourself: "*Does this response actually address the question that was asked?*" I remember once having an executive in a major corporation tell me, "*Don't answer what they ask, answer what you know.*" One thing I always want people to remember is *not to rely on Maslow's theory for achieving self-actualization*. It is based on a lack of information about the steps leading to that goal, and therein lies *most people's misperception*. Maslow himself had to fail at two prior careers before he hit on psychology.

How to respond:

"*The new information stands as an important point yet I need to return to the first question about...*"

8. Appeal to Emotion: Feelings Over Facts

What it looks like:

Emotional manipulation replaces logical reasoning as the method of establishing a point.

Real-world examples:

- The visual display of suffering animals used to oppose all animal research includes medical research that benefits human health.

- Fear of crime drives people to support policies that lack proven effectiveness in crime prevention.

- Patriotic images function as a strategy to avoid discussing the real expenses and benefits of military funding.

Why it's problematic:

Emotions should not substitute evidence-based analysis in decision-making processes.

When emotions ARE relevant:

- This category includes discussions about values together with priority-setting and life-changing policy effects.

- The consideration of emotions becomes important when they signal important concerns that need to be addressed.

How to respond:

"I understand that this matter creates strong emotional responses. We need to examine which methods show the best potential for achieving our desired goals."

9. Hasty Generalization: Jumping to Conclusions

What it looks like:

Limited examples or experiences lead to the creation of broad generalizations.

Real-world examples:

- The three teachers I know maintain summer breaks as part of their job, so teaching seems uncomplicated.

- The food at that restaurant is permanently dangerous because my friend developed food poisoning from it.

- Young people of this generation are considered lazy because they constantly use their phones.

- My two exercise attempts resulted in no weight loss, so I conclude exercise has no value.

Why we do it:

Our natural tendency as pattern-recognizing machines sometimes leads us to identify non-existent patterns in things. Personal encounters create more vivid and enduring memories than statistical information.

How to avoid it:

How many examples serve as my foundation for this conclusion? Do these specific instances properly represent the entire population? I need to examine additional information that could impact my evaluation.

10. Tu Quoque: "You Do It Too"

What it looks like:

The practice of avoiding criticism through the use of accusations that the critic performs the same behavior.

Real-world examples:

- Your criticism of my spending habits lacks validity because you purchased an expensive car during the previous year.

- Americans should not claim authority over human rights matters because they experience police brutality issues.

- You state that I should exercise more, yet you refrain from gym attendance yourself.

Why it's fallacious:

Even if the criticism is hypocritical, that doesn't make it wrong. People can give good advice even if they don't always follow it perfectly.

When hypocrisy IS relevant:

• When evaluating someone's credibility or sincerity

• When the critic's behavior directly contradicts their stated principles

How to respond:

"You're right that I haven't been perfect in this area either. That doesn't change whether the underlying point is valid. Let's focus on what the best approach would be."

Fallacies in Different Contexts

Political Fallacies

Common patterns:

• Straw man: Misrepresenting opponents' positions to make them easier to attack

• False dilemma: Presenting complex issues as simple either/or choices

• Ad hominem: Attacking opponents' character instead of their policies

• Appeal to fear: Using worst-case scenarios to justify positions

Example analysis:

Claim: *"My opponent wants to defund education because they voted against the school bond."*

Fallacies: Straw man (misrepresenting the vote) and hasty generalization (one vote doesn't indicate overall education policy)

Better approach: *"Let's look at both candidates' complete education platforms and track records."*

Marketing and Sales Fallacies

Common patterns:

• Bandwagon: *"Join millions of satisfied customers."*

• Appeal to authority: Celebrity endorsements for products outside their expertise

• False dilemma: *"Buy now or miss out forever."*

• Appeal to fear: *"Without this insurance/security system/supplement, you're at risk."*

Example analysis:

Claim: *"Nine out of ten dentists recommend this toothpaste."*

Questions to ask: How many dentists were surveyed? What were the other options? Were they paid for their endorsement?

Workplace Fallacies

Common patterns:

• Appeal to tradition: *"We've always done it this way."* I was once told to bring my own pencil to work.

• Sunk cost: *"We've invested too much in this project to stop now."*

• Hasty generalization: *"That approach failed at another company, so it won't work here."*

• False dilemma: *"Either we work overtime or we'll lose the client."* At one company, nobody ever left before 7pm because we had to show we were really working.

Example analysis:

Claim: *"If we allow flexible work schedules, productivity will plummet."*

Fallacy: Slippery slope without evidence

Better approach: *"Let's look at research on flexible schedules and maybe pilot a small test program."*

Responding to Fallacies Constructively

Don't be the fallacy police:

Constantly pointing out logical errors *makes you annoying and defensive*. Instead, focus on moving conversations onto more productive ground.

Strategies for constructive response:

1. Redirect to evidence:

"*That's an interesting perspective. What evidence have you seen that supports that view*?"

2. Ask clarifying questions:

"*When you say [position X], what specifically do you mean by that*?"

3. Acknowledge valid concerns:

"*I can see why you're worried about [underlying issue]. Let's look at different ways to address that concern*."

4. Suggest broader perspective:

"*I wonder if there are other ways to look at this situation that we haven't considered*."

5. Find common ground:

"*It sounds like we both care about [shared value]. What approaches might serve that goal*?"

Building Your Fallacy Resistance

When you hear an argument that immediately makes you feel strongly (angry, excited, fearful), pause and ask: "What specific reasoning is behind this claim? Is my emotional reaction making me skip over logical evaluation?"

The Steel Man Exercise:

Instead of looking for flaws in arguments you disagree with, try to understand the strongest possible version of the opposing position. This helps you avoid straw-man thinking and engage with real disagreements.

The Source Separation:

Practice evaluating arguments separately from who's making them. Ask: "If someone I disliked made this exact same argument, would I still find it convincing? If someone I respected made the opposite argument, would that change my evaluation?"

When Good People Use Bad Logic

It's important to remember that using fallacious reasoning doesn't make someone stupid or dishonest. We all use logical shortcuts, especially when we're tired, stressed, or passionate about an issue. Common reasons for fallacious thinking:

• Time pressure: Quick decisions often rely on mental shortcuts

• Emotional investment: Strong feelings can override careful thinking

• Cognitive overload: When juggling many decisions, we use simpler reasoning

• Social pressure: Group dynamics can discourage careful analysis

• Habit: We often copy reasoning patterns we've heard before

Approaching fallacies with empathy:

• Assume good intentions until proven otherwise

• Focus on the issue, not the person's intelligence

• Acknowledge when you catch yourself using fallacious reasoning

• Remember that everyone's thinking improves with practice

Chapter 6 Skills Checkpoint

Before moving on, ensure you can:

Recognize the top 10 fallacies in real-world arguments

Understand the psychological elements behind fallacies and their effectiveness

Constructive methods of handling fallacious arguments with professionalism

Implement knowledge of fallacies to assess arguments encountered in politics and marketing along with everyday life

Recognize the need to practice intellectual humility when you suspect your reasoning might include fallacious elements.

Self-Check Questions:

1. Which fallacy do you find yourself using most often? In what situations does this tend to happen?

2. Describe a time when you were convinced by an argument that you now recognize contained logical fallacies. What made it persuasive at the time? Give yourself some time on this one because you may have a bit of difficulty here.

3. How can you address fallacious reasoning in important relationships without damaging those relationships?

Real-World Application Challenge

Devote the next week to detecting fallacies within all your information consumption sources, including news articles and social media posts and advertisements and workplace discussions as well as family conversations. Practice the following steps:

1. Identifying the fallacy to yourself

2. Understanding why it might seem convincing

3. Thinking about what stronger evidence or reasoning would look like

4. Redirect discussions through appropriate inquiries which aim at better reasoning

The main objective is to shield yourself from deception by weak reasoning and to facilitate conversations based on evidence. Your enhanced ability to detect flawed reasoning now prepares you for handling complicated, uncertain situations that lack perfect information.

The next challenge awaits you to integrate all your acquired thinking abilities for making improved decisions in complex situations where complete information is unavailable.

Chapter 7:
Making Decisions in Uncertain Times

S he completely felt overwhelmed. The promotion included a 30% raise and future career growth. But there were also multiple things in her lifestyle that would need adjustment, abandonment, or retooling to meet the demands of this job.

For weeks, Maria compiled lists of advantages and disadvantages while she searched the internet for information about decision-making strategies and sought advice from everyone she knew. The accumulation of information created more uncertainty than clarity for her. Every advantage included a matching drawback in its structure. Every expert had a different framework. Each friend provided his or her own decision based on their personal life journey. It was discouraging. Where could she turn?

A mentor provided Maria with an essential perspective that completely altered her view: *"You attempt to decide with absolute clarity, but life presents us with uncertainty."* The true challenge isn't about removing uncertainty; it's about making sound decisions in an uncertain environment. How did she ever become so enmeshed in the idea that there was absolute certainty about anything?

Her new understanding transformed Maria's approach to decision-making. She concentrated on implementing a structured decision-making system that would enable her to choose the best possible solution from available information rather than pursuing the perfect decision. **There was no "perfect" solution.** When Maria moved to her new city with her family doing well, she discovered that becoming proficient at making choices in uncertain situations had become one of the most essential abilities she had ever acquired. And she got it simply by selecting the right mentor.

Traditional decision-making methods prove ineffective when dealing with complex situations. We need to get out of our comfort zone and explore new ways of approaching challenges.

We learn to make decisions by using basic choices like selecting vanilla or chocolate ice cream and choosing between two different colleges or two different jobs. *Real-world choices rarely provide straightforward options.* The beauty of the real world is contained in this complexity and diversity in almost every aspect of our lives.

Modern decision-making challenges:

- The abundance of factors to analyze presents a major challenge.

- Uncertain outcomes means you can't predict exactly what will happen

- Your goals exist in multiple dimensions because different results hold different levels of significance to you (money, family, happiness, security)

- You cannot spend unlimited time researching.

- The situation transforms during the time you need to make your choice.

- Your emotions regarding the decision strongly affect your ability to think clearly.

Why simple pro/con lists often fail

Jennifer prepared a list of pros and cons to decide about leaving her job to start her own consulting business.

Pros: Freedom, higher income potential, following her passion

Cons: Financial risk, no benefits, uncertain income, possible bankruptcy

The list failed to help her because it treated each factor equally important without considering the uncertainties involved. "Higher income potential" could mean anything from poverty to wealth. Her financial situation and other conditions determine whether "Financial risk" will result in minor or catastrophic outcomes.

A Better Framework for Complex Decisions

Step 1: Clarify What You're Really Trying to Accomplish

Your first step should be to *establish what you want to achieve* before you explore available options.

The "Why" Questions:

- What specific achievement do you want your choice to deliver?

- What are the essential elements that require attention in this particular situation?

- How would you define success during the upcoming year and the following five years? What's the plan?

- What core values hold the most importance in my current situation?

Maria's promotion example:

Initial framing: *"Should I take this promotion?"*

Deeper questions: What am I trying to accomplish in my career and life? How do I balance professional growth and family stability? What does success mean to me at this stage of my life?

Maria's insights: She understood that career advancement was important to her while she wanted to preserve her family connections. Success represented both professional advancement together with maintaining family contentment and strong bonds.

Step 2: Generate Options (More Than Just A or B)

People tend to evaluate insufficient choices because they automatically select either doing it or not doing it. Creative option generation often reveals better paths. Begin a list of potential options that may be more viable. Don't just start cold.

Option generation techniques:

The "What Else?" Method:

Force yourself to *create at least three additional options* following your initial two choices regardless of their probability.

The "Combining" Approach:

Search for opportunities that unite different option components.

The "Timing" Variation:

Look at the same fundamental choice through various time perspectives and different duration options.

Maria's expanded options:

- Take the promotion as offered

- Decline the promotion

- Negotiate to delay the start date by six months

- Ask about remote work arrangements for part of the week

- Take the promotion for two years with the agreement to return to the current city

- Propose a different role in the current city with some of the same responsibilities

- Use the offer to negotiate improvements in current position

Step 3: Evaluate Options Under Uncertainty

Since you can't predict the future perfectly, you need tools for thinking about decisions when outcomes are uncertain.

Scenario Planning: Thinking Through "**What If?**"

The process involves assessing multiple potential outcomes to determine how your choice would turn out in each scenario.

Basic scenario framework:

• Best case: Things go better than expected

• In the most likely scenario, events will unfold as you expect, but this may only last for a limited duration.

• Worst case: Things go worse than expected

• Unexpected case: Something happens that you didn't anticipate

Maria's scenario analysis for taking the promotion:

Best case: Family adjusts quickly, kids love new schools, career advancement leads to even better opportunities, parents visit regularly and enjoy travel

Most likely case: Adjustment takes 6-12 months, some challenges with kids' transition, career benefits as expected, occasional guilt about distance from parents

Worst case: Kids struggle significantly with move, marriage stress from relocation, parents' health declines and she can't help, new role doesn't work out as planned

Unexpected case: Remote work technology advances make location less important, economic downturn affects new position, family discovers they love travel and adventure

Expected Value Thinking (Simplified)

Basic expected value concepts can be applied without requiring mathematical expertise.

Simple approach:

For each scenario, estimate:

1. How likely is this scenario? (rough percentage)

2. How good or bad would this outcome be for you? (scale of 1-10)

Maria's rough calculation:

Best case: 20% chance, value = 9

Most likely: 50% chance, value = 7

Worst case: 25% chance, value = 3

Unexpected: 5% chance, value = varies

The rough assessment indicated that accepting the promotion would yield a positive outcome in 70% of cases (7 or higher) which helped Maria assess her readiness for the 25% chance of major obstacles.

Step 4: Address Your Decision-Making Biases

Common biases in complex decisions:

- Status Quo Bias: Overvaluing current situations simply because they're familiar.

- Question to ask: *"If I were starting fresh today, would I choose my current situation?"*

- Sunk Cost Fallacy: Continuing courses of action because of past investments rather than future potential.

- The question should be: *"If I ignored my past investments, which path would lead to the most beneficial outcome?"*

- Overconfidence: Assuming you can predict outcomes more accurately than you actually can.

The evaluation should include: **"Are there any possible risks which I have not considered**?" What possibilities exist that my control abilities might be exaggerated?

Loss Aversion: Overweighting potential losses compared to equivalent potential gains.

The question to ask is: *"Do I focus more on potential losses than potential gains?"*

Confirmation Bias: Seeking information that supports your preferred option.

The question to ask is: *"What evidence would show that my preferred choice is incorrect?"*

Tools for Better Decision-Making

The 10-10-10 Rule (Welch)

Consider how this choice will affect your emotions at three different points: *ten minutes, ten months, and ten years.*

This helps you balance immediate emotions with long-term consequences.

Example:

- Taking the promotion brings both fear and enthusiasm within ten minutes.

- Taking the promotion will bring satisfaction if it succeeds but stress and regret if it fails.

- The career growth will become more important than the initial transition difficulties after ten years.

The Regret Minimization Framework

The well-known business leader Jeff Bezos used this method to establish Amazon. He envisioned himself at 80 years old to determine whether attempting and failing would bring him greater regret than staying idle.

Complex decision-making calls for the following approach:

- What option would cause the most regret in the future?

- Which alternative creates the highest possibility that you will ponder 'what if?' after making your choice?

- Even when an option may not turn out perfectly, I should choose the one that respects my core values.

The Advisor Perspective

Reflect on the guidance you would provide to your best friend or your child in this particular circumstance.

The approach helps you detach from emotional involvement to observe the situation with clearer perspective.

The Reversibility Test

You should determine how quickly you can shift direction if this option does not achieve its goals.

You need less certainty for making decisions when the path to reversal is straightforward. Decisions that prove challenging to change need more thorough evaluation before making a choice.

Maria's reversibility analysis:

- Taking promotion: Somewhat reversible—could potentially return to previous city within 1-2 years, though not to exact same position

- Not taking promotion: Moderately reversible—similar opportunities might arise again, but career momentum could be lost

Time-limited Decisions Require Special Consideration

When you can't research forever:

- Set decision deadlines: Set a particular time frame for decision-making, and then you should choose an option.

- Focus your research on the two to three essential pieces of information needed for decision-making.

- Choose "good enough" options instead of seeking maximum perfection during the decision-making process.

- Most decisions do not require absolute certainty to achieve good outcomes.

Real-world example:

Tom needed seven days to choose between the job offer and other options. He dedicated his research time to answering three essential questions.

- Are the job responsibilities aligned with his career goals?

- Is the company culture a reasonable fit?

- Is the compensation fair for his experience level?

Tom dedicated his time to two days of detailed investigation of specific areas rather than attempting to study every aspect of the company and role.

The Process of Obtaining Guidance from Other People Regarding Your Decisions

When to seek input:

- Decisions that affect other people significantly
- Areas where others have more expertise than you

When emotional overload occurs, people should seek additional perspective because it helps them manage their situation better.

Complex situations where multiple viewpoints are valuable

Be specific about what you want:

- *"I'd like you to help me think through this decision."*
- *"I've already decided—I just need emotional support."*
- *"I'd like your perspective on the risks I might not be seeing."*

Explain to them the core values you possess along with the constraints you operate within. The advice you receive becomes more effective when you provide clear details about your values and boundaries.

Ask about their reasoning:

- People should explain their thinking instead of providing unexplained opinions.

- Examine the source of each piece of advice.

- Base your decision on the person's experience and their ability to grasp your circumstances when seeking their advice.

- Avoid decision by committee.

- Request input from others yet understand that you bear the final responsibility for your decisions.

Decision-Making in Different Life Areas

Career Decisions

Framework questions:

- The decision should support my future professional objectives.

- Which abilities will I acquire or lose through this experience?

- This choice will influence my way of life together with my interpersonal connections.

- What's my backup plan if this doesn't work out?

- The decision between risk-taking and playing it safe makes me wonder which one I would regret more.

Example: Starting a business

Consider the following factors instead of simply deciding between yes or no:

- Begin your business while continuing to work part-time in your current role.

- Create a business partnership with someone who brings different abilities to the table.

- Start by testing the market through consulting and freelancing work.

- Save specific amount before making the leap.

- Set specific conditions to determine when you should adjust your path or go back to employment.

Financial Decisions

Framework questions:

- The choice supports my complete financial objectives.

- What alternative opportunities could I utilize instead of this financial choice?

- This choice would impact how secure I am financially.

- The decision I make stems from fear or greed or is it something else?

The framework requires you to describe your emergency strategy when investments or purchases fail to deliver their predicted results.

Example: Buying vs. renting a home

Consider scenarios like:

- Job market changes requiring relocation.

- The interest rate fluctuations impact how much a house costs.

- The expenses for maintenance exceeded the expected amount.

- The property value decreases within the region.

- Family dynamics shift leading to different needs regarding living space.

Relationship Decisions

Framework questions:

- I need to determine my essential requirements for this relationship at this moment.

- The assessment of our shared long-term targets.

- The identification of recurring patterns in behavior that can either persist or transform.

- The process of finding a suitable balance between my values requires me to maintain them while making concessions.

- The internal signal within me emerges without overthinking.

- Example: Marriage timing.

The question becomes what specific issues do we need to resolve first?

- How do we handle conflict and stress together?

- Are we growing in compatible directions?

- We need to define what "ready" means specifically to our relationship.

Learning from Your Decisions

The Decision Journal

The record includes essential decisions that require attention.

This document contains both your chosen path and the reasons behind it, along with details about your decision sources and predicted results and actual results. The assessment of real-world outcomes should include your predictions and observed results. You'll learn essential knowledge to help you improve your choices in the future.

Regular review:

- Every six months, review your decision journal to identify patterns:

- When are your most successful decisions made?

- The judging process that affects your decision-making ability occurs because of what biases consistently appear?

- The accuracy rate of your predictions regarding results stands at what level?

- What decision-making tools serve you best?

The "No Regrets" Perspective

The assessment of past decisions should occur through evaluation of the decision-making process rather than assessment of final results because these outcomes remain beyond your control. I have used appropriate information to form my conclusion.

- Have I applied the best reasoning available to me?

- Did I consider my values and priorities?

- I have assessed the various possible scenarios.

- I have requested suitable feedback from people who are knowledgeable about this subject.

Good decision-making process leads to valid decision outcomes regardless of the eventual results. The success of an ill-conceived decision does not transform it into a good decision because good luck can occur.

Chapter 7 Skills Checkpoint

The following items need confirmation before moving forward:

Clarify decision goals: You should define the actual purpose behind your current choice which extends beyond immediate choices.

The generation of multiple options represents a shift from basic either/or choices towards innovative possibilities.

The ability to think in scenarios allows you to analyze different potential outcomes of your decisions.

The ability to identify decision biases includes recognizing how status quo bias and sunk costs, along with other biases, shape your judgment.

Use decision tools: Apply frameworks like 10-10-10, regret minimization, and reversibility tests

Learn from outcomes: Evaluate your decision-making process rather than just outcomes

Self-Check Questions:

1. A complex decision from the past challenges you. The frameworks presented in this chapter would have helped you approach this decision differently.

2. Which decision-making biases do you think affect you most often? In what types of situations?

3. You faced a decision-making dilemma because of uncertainty at a particular moment. Scenario thinking could have provided you with beneficial assistance during this situation.

Real-World Application Challenge:

A crucial decision exists either currently or will emerge in the near future. Use the comprehensive decision-making approach described in this chapter.

- Clarify your underlying goals and values.

- At least five different solution choices need to be produced.

- Analyze the potential results of your main alternatives by developing scenarios for the optimal and worst-case and most likely situations.

- Check for decision-making biases.

- At least two decision-making tools (10-10-10, regret minimization, etc.) should be implemented in your analysis.

- When suitable, seek guidance from others following the established procedures.

Create a document of both *your process and your decision*. Check how the situation developed along with your insights about decision-making processes during a six-month review.

The objective is not perfect decision-making but rather making decisions with available information and time. Like Maria did when she realized that family and career objectives can coexist through intentional choice-making, you will build self-assurance to handle ambiguity instead of being frozen by it.

Your enhanced decision-making abilities prepare you to effectively share your thought processes with others while participating in discussions that embrace diverse perspectives in a respectful and constructive manner.

Chapter 8: Communicating Your Critical Analysis

The management team received a solution from David which he considered outstanding for their customer service issues. He invested time in data analysis to create a plan that promised both time reductions of 40% and financial savings. He requested the next management meeting for the purpose of presenting his ideas.

David observed his audience members staring blankly at him when he reached the fifteen-minute mark during his presentation at the conference table. After finishing his presentation, his boss responded by saying, "*That's... a lot of information, David. Can you send us a summary?*" During the presentation, the CFO appeared perplexed while asking three questions that David had already explained in his presentation. His solution failed to receive approval because his presentation lacked clarity rather than because his solution was incorrect.

David observed his colleague receive approval for a comparable yet less detailed proposal at a conference meeting, that occurred six months after his initial attempt. The difference? Through learning, he mastered *the art of simplifying complex concepts into understandable messages*, which motivated people to take action. How did he do it?

David learned that his exceptional critical thinking abilities became pointless because he could not effectively convey his understanding to others in an understandable manner. His rightness became meaningless because he needed to discover methods to make others see his perspective.

The main factor that prevents smart analysis from persuading people becomes evident.

The Curse of Knowledge

Knowledge penetration *leads to forgetting* what it was like *being uninformed* about the subject. People who possess your level of understanding often miss out on information that others find hard to grasp. A college professor advises the test-taking student to "*Write in a way that helps the reader understand what you already know. You are writing for someone who is uninformed. Inform them.*" Looking over what she had written, the student realized that anyone without knowledge of this subject wouldn't easily understand her points.

Information Overload

A comprehensive research effort drives you to present every piece of discovered information. People become disinterested by excessive information delivery, which makes them believe topics become too complex to understand. Learn to simplify and drill down to the absolute necessary basics. Anyone writing flash fiction stories knows this from the get-go. We want to avoid the boring factor here. How do you know when people are bored? They start squirming in their seats, pulling out their phones, or checking their watches.

Missing the Human Element

The method of critical analysis tends to focus on evidence and logic but disregards the emotional and relationship factors and operational restrictions that impact your audience. Emotion is essential, but present it in an acceptable way. Draw your audience toward you. I do suppose many people start their presentations with a joke? They want to loosen up the audience and gain a degree of being liked at the beginning.

Wrong Level of Detail

The presentation of technical details to audiences who require general knowledge or the presentation of abstract concepts to those who need concrete specifics happens. Remember, you are building a house—not literally, but mentally and you need to give the workers what they must have for the final construction. Thinking of it this way makes it much simpler to do.

Forgetting the "So What?"

The explanation of your findings lacks clarity about how they relate to your audience. Keep a checklist for yourself handy.

Understanding your audience is all-important because everything you do will be fine-tuned to meet their needs. How do you construct a really impressive presentation?

Before crafting any communication, ask:

- What does my audience already know?

- What background information do they have?

- What misconceptions might they hold?

- What previous experiences shape their perspective?

What do they care about?

- What are their goals and priorities? How did you locate this information?

- What keeps them up at night? You want to keep them thinking about this long into the night.

- What would motivate them to act? What's the appeal that will get to them?

- How do they prefer to receive information? Depending on preference, you might want to use charts rather than videos, or vice versa.

The information delivery methods preferred by your audience include either detailed explanations or general overviews.

- Do they like data and details, or big picture concepts? The preferred learning method of visual learners includes using charts and diagrams.

- Do they prefer written analysis or verbal discussion?

- What constraints are they working under? How much time do they have?

- What approval do they need from others?

- What resources are available to them?

Real-world example:

Sarah attempted to convince her family to shift their cell phone service for financial savings. Her first approach involved detailed cost breakdowns and technical comparisons. After her first failure, she understood her family valued convenience above all else, including reliability more than cost reduction. She changed the focus of her analysis to presenting "getting the same service at reduced hassle and lower bills," which proved more effective.

The SCRAP (Sridharan) Framework offers a structured method to build arguments for maximum effectiveness

S - Situation: Establish the context and why this matters

C - Complication: Explain the problem or opportunity

R - Resolution: Present your analysis and recommendation

A - Action: Specify what you want people to do

P - Payoff: Explain the benefits of taking action

Example: David's customer service presentation was now revised according to the steps he needed to take to have it receive more attention, shown as an opportunity. have greater of resolution. appeal, and fit their needs. It outlined customer service times, reduction in response time, a pilot program, and lower costs. Everything fits within a more pleasing, action-oriented presentation for him.

The Pyramid Principle (Minto)

- **Begin with your conclusion** before providing evidence that follows an importance-based organization.

- Start by guiding your audience through their entire discovery process without direction.

- Try: "*The analysis shows that I support X because of three key elements A, B and C. I will explain each point separately.*"

Benefits:
- The primary message reaches busy individuals right away.

- Details-oriented viewers can watch the supporting evidence portion.

- You prevent others from becoming lost in analytical details

The Story Arc Method

Every writer who either takes a course at school, attends a writing seminar, or reads a book on the craft of writing knows about the story arc. *Humans have an innate tendency to follow organized sequences of events.* The structure of your analysis should resemble a story with these three main parts:

- The analysis begins by *showing the present circumstances* and the need for transformation. You're laying out the plan that you are then going to refine and display at length.

- Middle: The journey of analysis and *what you discovered*. For writers, the book would be "The Hero's Journey" or "Save the Cat (Screenwriting).

- The conclusion brings together *the solution* and its following actions.

Example:

Six months back we detected that our leading customers were paying their invoices with extended durations [**beginning**]. The investigation of payment delays became my focus [**middle transition**]. The results of my investigation brought unexpected findings to light. [**middle content**]. The proposed solution proves simpler than our initial assessment [**end transition**]. I propose these actions as my final recommendation [**end content**].

Making Complex Ideas Accessible

The Analogy Bridge

Familiar concepts help explain concepts that people do not understand well. Use what they already know and help them to understand different material that may be new or complex to them.

Weak explanation: *"Our database architecture creates redundancy issues that compromise data integrity."* This is one where people will want to take a quick nap.

The database functions like a library which stores identical books in various locations yet librarians sometimes fail to update multiple copies when they modify one. Different versions of information appear to customers based on the specific document they encounter.

It's always been my belief that if you truly understand something, you can explain it in basic terms so that someone else can understand it, too. If you can't do that, you need to go back to the sources and try to re-educate yourself so that you have a better understanding. *One highly recommended way of learning anything is studying it and then explaining it to someone else* **who knows nothing about it**. I remember walking around our university before a major test. Each of us asked the other to explain something. My test, as I recall, was about a small plant called the liverwort that lacks true roots. I was able to explain it to someone in great detail and I knew that I knew the material.

In the past at one of my jobs where we used computers, I had to explain to a physician about operating systems and how programs integrate with them. (Now I'm becoming familiar with "wraparound programs" or "wrappers" and apis.) Using the simplest example I could, I outlined the *operating system as building a foundation for an office complex* and the programs as the *furnishings within the buildings*. It was simple; it was straightforward, and he understood it very well.

The Layered Explanation

Present your idea at three levels of detail:

- Headline level: One sentence summary

- Executive summary level: Key points in 2-3 minutes

- Deep dive level: Full analysis for those who want it

Example:

The modification of our return process allows us to decrease customer complaints by 50%.

The main reason for complaints arises from our complicated return procedure, which leads to both customer dissatisfaction and staff exhaustion.

Executive summary: The majority of complaints stem from our complicated return process. This streamlined system enhances customer satisfaction and decreases employee workloads.

Deep dive: Detailed analysis of current process, comparison with industry best practices, implementation timeline, cost-benefit analysis, etc.

Visual Communication

A well-designed chart alongside diagrams often conveys information better than written paragraphs do. We've included a section later in this book that will show you how to prepare mind maps and where you can get them. If you want to skip to that right now, I highly recommend mind maps as a way of outlining and helping yourself to understand something.

Good visuals:

The key point stands out in the visual content instead of just displaying data.

The explanation needs to be easy to grasp right away.

Support your argument rather than replace it

Example visual approaches:

- Before/after comparisons
- Process flowcharts
- Simple trend lines
- Cost/benefit charts

Handling Disagreement and Objections

The Steel Man Approach

You should confront the most robust version of opposing positions instead of weakening them into easier targets for attack.

Weak approach: "*Some people think this is too expensive, but they're not considering the long-term benefits.*"

Better approach: "*The most thorough critique of this approach suggests that initial costs remain high while the expected advantages remain uncertain. I'll show why this investment holds value even when facing these genuine concerns...*"

The Acknowledgment Bridge

The presentation of your counterargument should begin with a recognition of valid concerns.

The method consists of two steps: "*I understand [concern] because [validation]. At the same time, [your perspective] because [evidence].*"

"*The process of changing our current approach faces legitimate risks together with adjustment expenses which lead to process disruption. The monthly customer loss from our current approach makes the need to change more critical than the costs of altering it.*"

The Question Invitation

You should avoid predicting every challenge so that you can encourage others to ask questions. Questioning is the beginning of getting your audience to relate to your material.

Effective phrases:

- *The potential drawbacks of the approach should be explained with the following question: 'What concerns would you have about this approach?'*

- You need to verify these conditions for the solution to work in your location.

- What elements from your perspective have I failed to notice?

- The analysis contains several questions that need clarification from the audience.

Different Communication Contexts

Written Communication

Email summaries:

- Start by presenting the main finding or recommendation.

- Multiple items benefit from being presented through bullet points.

- A one-sentence explanation should be included to demonstrate why the information holds importance.

- The report details exactly what actions require execution.

Formal reports:

- Executive summary for decision-makers.

- The methodology for implementers should be detailed in the report.

- All supporting data should be placed in easy-to-access appendices.

- The recommendations contain a precise outline of responsibilities for each party and their corresponding deadlines.

Informal updates:

- Keep your focus on changes which have occurred since the previous communication.

- Highlights any choices that need to be made.

- Use an organized structure to enhance the conversational tone.

Formal presentations:

- A presentation begins with both an agenda list and a time schedule.

- Signal transitions clearly (*"Now let's look at..."*).

- The speaker should verify understanding from the audience throughout the presentation.

- The presentation should conclude with defined next steps.

Meeting discussions:

- Start by delivering a brief summary of your stance using one or two sentences.

- Pause for important points to absorb into listeners' minds. Keep telling yourself not to rush.

- Use past collaborative work together or decisions from previous times.

- Specific feedback requests should be directed to you instead of seeking general reactions.

One-on-one conversations:

- Adjust your speaking style to match the communication style of the person you are speaking with.

- More dialogue with less monologue should be your communication method.

- Check understanding by asking questions instead of only nodding your head.

- You should be ready to either explore the topic in depth or maintain a general level depending on their level of interest.

Building Consensus Through Critical Dialogue

Establish ground rules:

- The discussion should concentrate on concepts instead of targeting personal traits.

- Before you can seek understanding from others, you need to first understand their perspective.

- The analysis of assumptions serves as the main focus, while the analysis of motivations should be avoided.

- Each person should try to find opportunities to enhance the concepts of their colleagues.

Use inclusive language:

- Instead of using "*You should*" the phrase "*What if we*" becomes a better option. In psychology, we call this re-framing.

- I am puzzled about whether the following statement is correct: "*I'm wondering whether...*" rather than "*Obviously...*".

- *"I need clarification regarding this topic, but your opinion is incorrect about it,"* so you ask them for an explanation.

The Collaborative Analysis Approach

- Create an open dialogue that invites others to join the thought process before presenting any conclusions.

- The traditional method presents results first which is followed by a recommendation.

- I have examined this problem through analysis and my findings are as follows: *What patterns do you see? What patterns do you recognize and what elements do I fail to detect?*

The Collaborative Analysis Approach enables the following benefits:

- When others take part in creating the conclusion, they develop ownership over it.

- Your analysis will receive valuable feedback, which enhances its overall quality.

- The involvement of others in creating solutions reduces their opposition toward the solutions.

Finding Common Ground

When people disagree about solutions, they typically maintain common goals or concerns.

Example:

- Person A attempts to reduce costs by implementing staff reductions.

- The maintenance of service quality remains a priority for Person B.

- The organization's financial stability and customer service quality represent the common objectives of both parties.

- A better question could be, "How do we achieve both superior service quality and operational efficiency?"

Practical Communication Exercises

The Elevator Test

Develop your essential findings into *brief summaries that last 30 seconds* and extend to 2 minutes and 10 minutes. The content should be comprehensive while its length matches the allocated time frame. This type of pitch is often used while people are literally in an elevator, where there is extremely limited time to get an idea and implications across.

The Audience Switch

You should present the same analysis to different groups, including:

- You should present the business-oriented impact of the findings to your boss.

- Your presentation should demonstrate how to overcome the difficulties of implementation to your peers.

- Your presentation should explain what changes will affect your team members.

- When presenting to a doubtful colleague, you should concentrate on showing how their doubts will be resolved.

The Objection Anticipation

Before important presentations or discussions:

- List the top 5 objections you might hear

- Prepare thoughtful responses to each

- Consider how to acknowledge valid concerns

- Practice responding without defensiveness

The Clarity Check

After explaining something complex:

- Ask specific questions for clarity: *"What questions do you have about the timeline?"* rather than *"Does this make sense?"*

- Request the person to restate the main recommendation they heard.

- Check for buy-in: *"What parts of this approach feel right to you? What concerns do you have?"*

- The communication techniques of the digital era need consideration.

Virtual Presentations

Engagement strategies:

- The implementation of polling tools or chat functions helps participants become more involved.

- Your presentation should include strategic screen display followed by hiding the content.

- You should integrate more processing time during your presentation.

- You need to check in on the participants more often than traditional in-person meetings.

Technical considerations:

- Test your setup in advance.

- All technical issues need a backup solution.

- The use of larger fonts along with simple visuals should be implemented.

- The audience has limited time, so the information needs to be delivered quickly.

- Asynchronous Communication

When documenting analysis:

- You must include your thought process and reasoning in addition to your conclusions.

- The document should feature headings together with bullet points to help readers quickly understand it.

- The information should be presented at three different levels to accommodate different reader needs.

- Include contact information for questions

Information Overload Management

- The main points should be summarized at the beginning of the document.

- Use attachments for detailed supporting data.

- The distinction between action requirements and informational content should be clear.

- You must specify the time frame for receiving responses.

Chapter 8 Skills Checkpoint

You need to confirm your ability to do the following tasks before proceeding:

Adapt to your audience: Tailor your communication style and content to your audience's needs, knowledge level, and preferences

Structure arguments effectively: Use frameworks like **SCRAP and the Pyramid Principle** to organize your thinking for maximum impact

Make complex ideas accessible: Use analogies, layered explanations, and visuals to help others understand sophisticated analysis

Handle disagreement constructively: Engage with strong opposing arguments and build consensus through dialogue

Communicate across different formats: Adapt your approach for written reports, presentations, meetings, and one-on-one conversations

Self-Check Questions:

1. Consider a situation in which your excellent analysis failed to influence others. What communication elements may have led to this outcome?

2. Among your target groups consisting of bosses and peers and skeptics and others, which ones do you find most difficult to communicate with? Why?

3. Which particular method from this chapter would improve your following crucial presentation or dialogue?

Real-World Application Challenge:

You need to communicate a complex idea or recommendation to an audience within the upcoming month. Use the frameworks presented in this chapter:

- Examine your audience by answering the provided questions

- Develop your argument by using SCRAP or the Pyramid Principle

- Create visual aids or analogies to simplify complex information.

- The three main counterarguments need identification for developing effective answers.

- Practice presenting your main point for 30 seconds, 2 minutes and 10 minutes.

- After your communication, reflect on what worked well and what you'd do differently

- Use the "acknowledgment bridge" method in your upcoming disagreements or challenging discussions.

The purpose of communication should be to present your ideas so others can comprehend your perspective and interact meaningfully with your thoughts. You can develop the necessary skills for critical thinking implementation through your efforts just like David learned about the importance of brilliant communication for brilliant analysis.

The current era presents unique challenges that you can navigate by utilizing your developed communication skills and your ability to work with AI tools, all while maintaining your independent critical judgment.

Chapter 9: Working with AI as a Thinking Partner

Lisa expressed doubts about her team's usage of ChatGPT for marketing campaign brainstorming when she first learned about it. She believed artificial intelligence systems lacked the capability to understand customers like humans do. Her perception about artificial intelligence changed after she observed her colleague create twenty creative concepts through ChatGPT in ten minutes, which otherwise would have taken their team multiple hours to develop. *"It's eye-opening,"* she thought.

Now, six months into her journey, Lisa made an impressive discovery: she realized that AI enhances her mental processes instead of replacing them. The technology at her disposal helped her produce first drafts of ideas as well as test new concepts and gather information about unfamiliar subjects before conducting real-life conversations. She gained vital lessons regarding AI misdirection and developed skills for trusting its outputs and maintaining independent critical thinking abilities. One of the lessons that she was to learn was about AI hallucinations, and that would open a whole new world of discovery for her. Also had to learn how to effectively use "prompting".

According to Lisa, the essential concept requires teams to understand how to interact with AI by leading when appropriate and following instructions at suitable times while keeping individual thinking capabilities intact. This learning curve wasn't steep, but it presented new challenges and a new way of thinking.

The fundamental transition in all industries and life aspects matches what Lisa experienced. Our society is entering a period where critical thinking with artificial intelligence becomes the most valuable skill. But there's a bias, and we need to work within that framework.

Understanding AI's Strengths and Blindspots

What AI excels at:

• Processing vast amounts of information quickly: AI can analyze thousands of documents, studies, or data points in seconds

• Generating multiple perspectives: It can help you see problems from angles you might not consider. Chatbots will even suggest additional questions you might ask in a prompt. Can lead you to a deeper understanding of the material.

• Identifying patterns: AI can spot trends and connections across large datasets. This makes it especially useful in medical research.

• Brainstorming and ideation: It can produce numerous creative options rapidly.

• Explaining complex topics: AI can break down difficult concepts into understandable pieces and provide helpful summaries.

• Role-playing scenarios: It can simulate different viewpoints for practice or analysis, such as what is called "sandboxing."

What AI struggles with:

• Understanding context and nuance: AI might miss cultural sub-tleties, organizational politics, or personal relationships that affect decisions

• Accessing current information: Many AI models may have knowledge cutoffs and can't access real-time information, and some don't have direct access to the internet.

• Evaluating emotional and interpersonal factors: AI can describe emotions but doesn't truly understand their impact on human decisions

• Making ethical judgments: While AI can explain ethical frameworks, it can't make nuanced moral decisions for specific situations.

• Understanding your unique circumstances: AI doesn't know your personal values, constraints, or long-term goals unless you *explicitly provide that context*. You can set rules for your writing style, voice, tone, and what to avoid. All of this, in certain chatbots, can be retained in memory for future use. Here's where prompting comes into play, and you must carefully outline what you need in that prompt. If you are unclear about prompting, I suggest you do a bit of research on it and perhaps watch some YouTube videos that explain it.

Real-world example:

Mark used AI to help evaluate job offers. The AI analysis delivered an outstanding evaluation of salary comparisons and industry trends, as well as career progression information. The system could not consider that Mark's elderly mother lived near his current workplace while his wife operated a business locally and Mark had learned from high-pressure work to value work-life balance above salary. Mark needed to combine AI analytical functions with his individual background and his personal values.

AI as a Research Assistant

Effective AI research strategies:

Start with **broad exploration**, then **narrow down.** Instead of asking, "*What's the best investment strategy,*" try:

• "*What are the main categories of investment strategies for someone in their 30s?*"

• "*What are the pros and cons of index fund investing versus active stock picking?*"

• "*What questions should I ask when choosing a financial advisor?*"
Ask for multiple perspectives:

• What opinions would a cautious investor have about cryptocurrency? What position would an investor who takes risks have regarding this topic? You can even ask for a brief explanation or summary of cryptocurrencies.

• What are the strongest arguments for and against remote work policies?

• How would various age groups react to a social media approach?
*Request evidence and sources (*these *are all prompt-related):*

• What research supports this recommendation?

• Where can I find more information about this topic?

• What limitations exist in the studies that you are using as references?

Real application example:

Jennifer examined whether an MBA degree was the right choice for her future. She used AI to gather strategic information instead of asking if she should pursue an MBA. Asking whether she should get a degree is asking for an opinion, and AI doesn't offer opinions but data and data summaries.

Research questions she asked AI:

The different MBA programs, together with their standard graduation outcomes, made up one of her research questions.

The decision to pursue an MBA required evaluation of which factors were most important.

She wanted to know what options existed for career growth in marketing that would be better than earning an MBA.

And she wanted to know which questions she should ask of MBA students who are currently enrolled and who have already graduated.

What she did with the AI's responses:

The information she got allowed her to choose interviewees, educational institutions for study, and criteria for decision-making. Although the AI system provided her with better research data, she chose her path after *speaking with people in various programs.*

AI for Enhanced Critical Thinking Through Prompting

Use AI to challenge your assumptions using reflective prompting techniques.

(*The following methods stem from educational practices that Jules White from Vanderbilt University has documented.*)

AI helps you question your beliefs through reflective prompting methods.

1. The Assumption Challenger

"*I'm considering [your position]. What are some compelling arguments that challenge this view? What hidden assumptions might I be relying on?*"

2. The Perspective Balancer

"*Here's what I believe: [your argument]. Present the most compelling counterargument to your position while explaining its potential effectiveness for an outside audience.*"

3. The Lens Changer

"*I'm analyzing [situation]. What additional perspectives from different cultures or professions or life experiences could I be overlooking?*"

Example in Practice:

Tom was enthusiastic about proposing a 4-day workweek policy at his company. Before finalizing his recommendations, he asked AI:

"Would a 4-day workweek really improve productivity and satisfaction? What objections might critics raise? What operational hurdles should I consider?"

AI returned insights about service disruptions, part-time scheduling hours, and conflicting study results across industries. While Tom still supported the policy, he refined his plan by proactively addressing anticipated objections.

4. The "What Happens If" Prompt

If [scenario] were to happen, what ripple effects might follow? What are a few plausible directions this situation could take?

5. The Risk Radar

What unexpected outcomes could occur because of [decision or strategy]? What blind spots am I not seeing?

6. The Possibility Probe

What fresh possibilities could emerge from [trend or development]? Who might benefit—and who might be left behind?

7. The Risk Assessor:

What negative elements could arise from [plan/decision]? What risks might I not be considering?

8. The Opportunity Spotter:

What possibilities could appear because of [situation/trend]? Which groups would gain advantages through these developments?

AI for Decision Support

AI requires information, and *you need to provide enough of it* for it to meet your prompt requests. Instead of saying, *"What should I do?"* provide context:

Step 1: At 35 years old and working as a marketing manager, I need to choose between the promotion that requires moving from Denver to Atlanta. The new position brings a 25% salary increase, yet forces me to depart from my support system. What factors should I consider?

(**The AI needs sufficient information to fulfill your request.**)

Step 2: Use AI for option generation

AI should generate all available options from this situation along with innovative choices that might escape human observation.

Step 3: Apply AI analysis to your specific context

How would you rate these options when I have [your values] as priorities and [your constraints] as my limitations?

Step 4: Stress-test your thinking

What mental biases might influence my decision-making process about this choice? Ask which questions will guide your selection before making this choice.

Step 5: Retain final judgment

AI should *function as one element* among others in your decision process *without becoming the deciding factor*.

Real-life example:

Sarah used AI to help decide whether to start a side business, and AI helped with the following:

- The analysis revealed 12 distinct business strategies that she could start.

- The document presented an organized plan for market research activities.

- The analysis presented information about the legal and tax consequences for business operations.

- The system produced multiple questions that would be useful for other entrepreneurs to ask.

Sarah maintained command over:

- The selection of her business concept that matched her professional talents and personal interests.

- The assessment of how much risk she should take on.

- She decided if the current time suited her household circumstances.

- AI provided advice that proved useful based on her individual circumstances.

- Maintaining Independent Judgment

Red flags for over-relying on AI:

You stop thinking for yourself:

• Taking AI recommendations without understanding the reasoning

• Not considering your unique circumstances

• Avoiding difficult decisions by deferring to AI

You trust AI more than human expertise:

• Choosing AI advice over qualified professionals in high-stakes situations

• Not seeking human input on complex personal or professional decisions

• Assuming AI has access to more current information than it actually does

You lose touch with your own values and intuition:

• Making decisions that feel wrong but that AI recommended

• Not considering emotional and relationship factors that AI can't fully understand

• Forgetting to factor in your personal goals and constraints

The Verification Habit involves strategies for maintaining judgment. For important AI-generated information, verify through other sources:

- Check recent news for current developments
- Consult human experts for specialized advice
- Test AI's reasoning against your experience and knowledge

The Context Check always requires that you ask yourself:

- Does AI understand my specific situation? The entire question of "understanding" in AI is **still hotly debated**.
- What important factors might AI be missing?
- How do my values and constraints affect this analysis?

The Gut Check means after getting AI input, pause and ask:

- Does this feel right given what I know about myself and my situation?
- What does my intuition say about this recommendation?
- What would I decide if I couldn't use AI?

AI Communication and Collaboration

Using AI prompts to improve your communication involves drafting and refining:

- Generate initial drafts of difficult emails, presentations, or proposals
- Ask AI to make your writing clearer or more persuasive. (There are specialized programs for helping with writing clarity and grammar.)
- Practice explaining complex ideas in simpler terms

Perspective taking:

- How might [specific person] respond to this message?
- What concerns might my audience have about this proposal?
- How can I address objections while remaining respectful?

Argument strengthening:

- How can I make this argument more convincing?
- What evidence would strengthen this position?
- Where are the weak points in my reasoning?

Example:

Before a tough conversation with her boss about work-life balance, Monica used AI to:

- Role-play the conversation from her boss's perspective
- Identify potential objections and prepare responses
- Practice explaining her position in different ways
- Generate specific proposals rather than just complaints

The AI practice session helped Monica engage in a more productive conversation with her boss. Had you ever thought of having a practice session with an algorithm? No, but you have had practice sessions either before a mirror or with a friend when you were going to make a presentation of some importance, didn't you? It's the 21st-century edition now and AI's going to be your practice partner. .

AI in Different Life Contexts

Professional Development:

AI can help with:

- Identifying skills to develop for career advancement
- Generating learning plans and resources
- Practicing job interview scenarios
- Analyzing industry trends and opportunities

You should still:

- Network with real professionals in your field
- Seek mentorship from experienced colleagues. There's no substitute for experience in a field, and anyone with it may be able to guide you more effectively than an AI algorithm. College courses may never fully prepare us for careers in any field, and as a psychologist, I know that from personal experience.

• Make career decisions based on your personal goals and values. Remember, *AI doesn't "understand" values, ethics, or morality,* so don't depend on it for any of those things. It can only look at data and give you the results from that analysis.

• Verify AI's information about specific companies or opportunities

Personal Finance:

AI can help with:

• Explaining investment concepts and strategies

• Analyzing different financial scenarios

• Generating budgeting frameworks

• Identifying questions to ask financial advisors

You should still:

• Consult qualified financial professionals for major decisions

• Consider your personal risk tolerance and life circumstances

• Verify AI's information about current market conditions

• Make final investment decisions based on your own research and professional advice

Health and Wellness:

AI can help with:

• Explaining medical conditions and treatment options, but only from currently existing data that it has access to on the internet. If it doesn't have access, it may indicate that it has to use material prior to its cutoff access date to the internet. This could lead to your getting outdated information, especially where it concerns health or medicine and current research findings. .

• Generating questions to ask healthcare providers

• Providing general wellness information

• Helping track symptoms or health metrics

You should still:

• Consult qualified healthcare professionals for medical decisions

• Not rely on AI for diagnosis or treatment recommendations. How many times have I heard a healthcare professional say, "*Oh, that's Dr. Google's recommendation?*"

• Consider your specific health history and circumstances

• Verify medical information through reputable health sources

The Future of Human-AI Collaboration

Emerging skills for the AI age:

AI Literacy:

AI literacy involves the ability to determine the range of capabilities and limitations of different AI tools as well as the operation of these tools and their proper usage scenarios. In effect, we're talking about tutorials or online classes. Just as only one book may not answer all of your questions sufficiently, one tutorial, course, or online seminar may still fail to provide what you need. Always look for more materials in your area of interest.

Prompt Engineering:

Learning to communicate effectively with AI systems to get useful responses.

Human-AI Team Management:

It requires the understanding of when to take the lead, when to accept the AI suggestions, and when to totally dismiss them.

Ethical AI Use:

This is the process of understanding the impact of AI decisions and maintaining human accountability for their use. Engineers of algorithms may try to insert ethics into the process, but it has been a difficult task. In fact, one of the biggest problems with algorithms in the 21st century is that there may be hidden biases in the coding that has yet to be discovered.

Meta-Critical Thinking:

Critical evaluation of the thinking process of AI, including the evaluation of its "reasoning" process, the identification of its biases, and the preservation of one's own judgment. Algorithms are really *compilations of computer programming patches* that have been cobbled together to make the algorithm. So, we must note that *bias is baked in*, and no one will know where it is, where it came from, or how to eliminate it. If you don't know what's there, how could you ask the necessary questions?

Stay informed about AI capabilities:
- Understand what new AI tools can do
- Learn about limitations and potential biases
- Keep up with best practices for AI use

Maintain uniquely human skills:
- Emotional intelligence and empathy
- Creative problem-solving
- Ethical reasoning and judgment
- Building relationships and trust
- Understanding context and nuance

Practice collaborative thinking:

Use AI as a *tool to enhance your thinking processes* instead of replacing them.
- Develop skills in directing and evaluating AI assistance
- Maintain your ability to think independently when needed

Practical Exercises for AI Partnership

Exercise 1: The AI Research Challenge

Choose a topic you need to learn about. Use AI to:

1. Generate an overview of key concepts

2. Identify important questions to research

3. Find different perspectives on controversial aspects

4. Create a learning plan

Then verify AI's information through human sources and evaluate what AI missed or got wrong.

Exercise 2: The Decision Support Practice

Use AI to help with a real decision you're facing:

1. Clearly define the decision and provide context

2. Ask AI to generate options you might not have considered

3. Use AI to identify potential risks and benefits

4. Ask AI what questions you should consider

5. Make your decision based on AI input plus your own judgment

6. Reflect on what AI added to your thinking and where human judgment was essential

Exercise 3: The Assumption Challenge

For one week, whenever you have a strong opinion about something, ask AI:

"I believe [your position]. What are the strongest arguments against this view? What assumptions might I be making?"

Notice how this affects your thinking and whether it helps you consider perspectives you hadn't thought of.

Chapter 9 Skills Checkpoint

Before moving on, ensure you can:

Understand AI capabilities and limitations: Know what AI can and cannot do well in different contexts

Use AI as a research tool: Leverage AI for information gathering while maintaining verification habits

Apply AI for critical thinking enhancement: Use AI to challenge assumptions, explore scenarios, and strengthen reasoning

Maintain independent judgment: Make final decisions based on AI input combined with your values, context, and human consultation

Collaborate effectively with AI: Know when to lead, when to follow, and when to ignore AI recommendations

Self-Check Questions:

1. In what areas of your life could AI assistance be most valuable? Where should you be most cautious about relying on AI?

2. How do you plan to verify important information that AI provides?

3. What strategies will you use to maintain your independent thinking while benefiting from AI capabilities?

Real-World Application Challenge:

For the next month, experiment with using AI as a thinking partner in one specific area of your life (work projects, personal decisions, learning new skills, etc.). Apply the frameworks from this chapter:

1. Clearly define what you want AI to help with

2. Use AI for research, perspective-taking, and option generation

3. Verify important information through other sources

4. Maintain your own judgment for final decisions

5. Reflect weekly on what AI added to your thinking and where human judgment was essential

Create documentation about your journey and modify your strategies through the knowledge you acquire about effective human-AI partnership.

The aim should be to develop a collaborative system where AI strengthens your cognitive processes, but you stay in charge of vital choices. Like Lisa, you can develop the skills to leverage artificial intelligence while strengthening rather than replacing your critical thinking abilities.

Your developed AI collaboration skills prepare you to create enduring practices that will help you preserve and improve critical thinking abilities regardless of technological or societal advancements in the future.

Chapter 10: Building Your Critical Thinking Habits

Rachel noticed her abilities had faded three years after finishing her MBA program. The business school had found her to be sharp with analytical skills while being confident when dealing with complex problems. The demands of her daily work, including meetings, emails, and deadlines, forced her to operate on autopilot, accept information without verification, and reuse established thought patterns from her past.

She shared with her study partner that she learned excellent frameworks and thinking tools during business school but never utilized them. But she was aware of how to think critically, yet she no longer used this ability in her daily life.

Her friend smiled knowingly. *"My professor called it '**skill decay**.' It's the term she taught me to describe this phenomenon. Critical thinking was a subject you studied, but you never turned it into a consistent* *practice. You possess all the necessary tools, but you need to establish automatic usage of them."*

The discussion motivated Rachel to view critical thinking through a new perspective. She started developing regular everyday practices, which helped her preserve her critical thinking abilities instead of saving these skills for major decisions. Colleagues noticed how her judgment improved as she asked better questions and handled problems with greater care.

Rachel learned an essential principle about critical thinking when she found that *this skill needs continuous practice* and development after initial learning.

Why Good Intentions Aren't Enough

The Knowledge-Action Gap

Critical thinking knowledge surpasses actual practice for most individuals. The understanding of confirmation bias or logical fallacies does not naturally transform your real-world thinking processes.

The Convenience Trap

Your mind will automatically use the simplest mental approaches whenever you face busy or stressful situations. Without established habits, you will use quick mental shortcuts to replace thorough analysis even when important decisions are at stake.

The Overconfidence Problem

Learning about critical thinking may create false beliefs that you already practice it effectively. Your increased confidence level actually diminishes your commitment to deliberate practice, which is essential for improvement.

The Context Challenge

Your critical thinking abilities remain strong when working on business projects, yet you completely give up critical thinking when handling health choices and political opinions. Critical thinking shouldn't be reserved for business because it is useful in every aspect of your life.

Real-world examples:

- A financial analyst who evaluates investment data at work but invests in get-rich-quick schemes outside of his professional role

- A scientist who needs evidence-based proof in her research but shares baseless health information through social media platforms

- The manager applies systematic decision processes to his business choices but makes spontaneous decisions when purchasing personally

The Habit Formation Framework for Critical Thinking

Understanding Habit Loops

Every habit follows a simple pattern: **Cue → Routine → Reward**. It's as sure as the sun will come up tomorrow, and you can depend on this routine being repeated.

For critical thinking habits:

- **Cue**: Something that triggers you to engage your thinking skills
- **Routine:** The specific thinking practice you do
- **Reward:** The benefit you get from better thinking

Example habit loop:

- **Cue**: When someone presents news which triggers strong emotions such as anger or excitement or worry

- **Routine**: You should ask yourself which investigations should come before taking any action or sharing this information.

- **Reward**: Being confident and informed while avoiding the embarrassment of sharing incorrect information

Daily Critical Thinking Practices

Morning Thinking Rituals

The Daily Assumption Check (2 minutes)

Each morning, choose an assumption you will make about the upcoming day, then *challenge its validity.*

- *"I'm assuming this meeting will be unproductive—what if I'm wrong?"*

- *"The evidence supporting my belief about my boss's dissatisfaction with my work remains unclear".*

- *"I'm assuming I don't have time to exercise—what if I looked at my schedule differently?"*

The Priority Reality Check (3 minutes)

Before starting your day, ask yourself these questions:

• Which daily activities hold more priority than their perceived importance?

• What important issue should I address despite my current avoidance?

• *"How will I know if today was successful?"* Note: How do you measure "success"?

The Information Diet Planning (2 minutes)

You should establish specific goals regarding how you want to use your information.

- What essential information do I need to understand during this day?

- You need to identify the sources of information you will use and those you will avoid.

- What amount of time will I dedicate to news and social

media?

Decision-Point Habits

The Pause Practice

- Before making important choices that exceed $50 or take more than thirty minutes of time or affect relationships, stop for 30 seconds.

- Before deciding on a course of action, ask yourself whether your thinking system matches the situation. What if it doesn't? What will you do?

- The second step involves considering what extra information would provide helpful value.

The Perspective Switch

When facing a problem or conflict:

- You need to understand how someone you respect would approach this specific situation.

- Consider: "*What would this look like from the other person's perspective?*"

- You should think about how someone else would handle this situation and how you would advise your best friend in the same position.

The Evidence Question

Before accepting or sharing information:

- The following question should be asked to verify the truth of this information: "*How do I know this is true?*"

- The source remains unknown to you while you need to figure out their motivations.

- You need to consider how much evidence would increase your confidence in this claim.

Weekly Reflection Practices
The Decision Review (10 minutes)
Every Sunday, review the week's significant decisions:

- You should make a list of all the major choices you made this week.

- Which ones used beneficial thinking processes?

- You should identify the decisions that would require different approaches if you encountered identical circumstances in the future.

- Do you recognize any specific patterns that appear in the decisions you make?

The Learning Audit (5 minutes)
Assess your intellectual growth:

- The week taught you important knowledge that transformed your understanding.

- You should identify new questions that you want to research in detail.

- What specific biases caused you to think differently about a situation?

- What assumptions did I question or revise?

The Relationship Check (5 minutes)

- You need to evaluate how your thinking methods affected your relationships with others.

- You need to identify the specific instances where effective questioning enhanced your weekly conversations.

- The way I heard others versus the way I waited to speak played a key role in this interaction.

- What disagreements did I handle constructively?

- The quality of your thinking patterns affected your social interactions with others.

Building Habits in Different Life Contexts
Workplace Critical Thinking Habits
Meeting Habits:

- You should prepare your questions for meetings before they start.

- The group makes certain assumptions when meeting but does not include every possible perspective.

- You need to establish the decided items and determine the upcoming steps after meetings.

Email and Communication Habits:

- You should take a 10-minute pause before responding to emails that create strong emotions. Working in psychiatric hospitals, I learned a lot about holding off on emotional reactions and standing back before responding to emails. Re-read the email and allow yourself some time to reconsider how written communications don't convey the true intent and may push an emotional reaction. Sometimes, speaking to someone is more effective than anything you could have written to them.

- You need to verify the accuracy of information before sharing it with others.

- Before sending: *"Is this clear, helpful, and necessary?"* Too often I've seen emails sent out by management that fail to include important factors that were involved in the sending of the email in the first place. You have to ask yourself, *"How did they miss that?"* There's one simple answer: proofreading.

Project and Problem-Solving Habits:
- Starting projects: *"What are we really trying to accomplish?"*

- At the middle point of the project we should evaluate what elements work along with what elements do not work and what adjustments need to be made.

- The conclusion of a project brings valuable learning that helps with upcoming professional tasks.

Financial Decision Habits
Daily Money Habits:
- The purchase of items greater than $20 should begin with an analysis of your motivation and exploration of alternative solutions which achieve similar results.

- When reviewing my investment accounts I need to understand my long-term financial plan and how short-term results match this strategy.

- You should determine how financial news affects your specific situation when you encounter it.

Monthly Financial Reviews:
- You should evaluate your spending by examining where your money went while making sure your financial decisions match your core values.

- Check your current status regarding financial targets along with determining what modifications you should implement.

- Your understanding of money during the month will serve as the main takeaway.

Health and Wellness Habits
Information Evaluation:
- Before implementing health recommendations you need to verify their origin and supporting evidence.

- The health news you read reveals whether it focuses on relationship patterns between elements or demonstrates the actual cause-effect connection between them.

- You need to understand what your doctor would think about your planned health modification.

Self-Awareness Practices
- A weekly energy audit requires you to evaluate your energy levels to determine when they peak and when they are lowest along with their regular patterns.

- The Monthly Habit Review allows you to evaluate which

health practices actually help you and which ones don't work.

- The quarterly goal assessment requires you to check if your health objectives match your values while being realistic.

Relationship and Social Habits

Communication Practices:

- I **perform three steps** before challenging conversations to determine my goals and effective communication methods.

- I will determine the protection goals and achievement targets of the opposing side during disagreements.

- The evaluation process after conflicts helps me understand both myself and the other person better.

Social Media Habits:

- The process of sharing requires evaluation to determine if the content is helpful and accurate and if it is truly needed.

- The evaluation of emotional responses from social media posts requires me to understand the reason behind the feelings and determine proper responses.

- Daily social media limit: Set specific **time boundaries** and stick to them

The Micro-Habit Approach:

- Begin your critical thinking development with activities which require less than thirty seconds to complete.

- A brief pause will occur between sharing news online.

- Ask one question before giving advice.

- Stop and evaluate your emotional response before you respond to emails.

The Integration Strategy:
- The implementation of critical thinking processes should integrate into daily procedures.

- The process of questioning assumptions should occur during your daily commute.

- The time spent during lunch breaks should be used for perspective-taking activities.

- Daily reflection occurs while performing the simple act of brushing your teeth.

The Time Investment Reframe:
- The process of good thinking produces time savings that extend into the future.

- The ability to make quality choices leads to fewer correction requirements.

- The right questions help prevent communication breakdowns.

- Careful analysis prevents costly mistakes

Environmental Cues:
- Your environment needs modification to **deliver reminders** about your schedule.

- A thinking **question image** should be set as your phone wallpaper.

- The **calendar** includes weekly review reminders.

- Place **sticky notes** in places where you will see them such as bathroom mirrors, computer screens and car dashboards.

Habit Stacking:

- You should connect your new thinking behaviors to your current established practices.

- My daily assumption check begins right after I pour my morning coffee.

- A review of daily priorities should be done before checking your email.

- During my car rides I reflect about the essential lesson I discovered throughout the day.

Start Small:

- Begin with a basic thinking habit that you can gradually increase over time.

- The first week involves asking *"How do I know this?"* **once per day.**

- Add daily assumption check to your routine during week two.

- The third week requires students to perform a weekly review of their decisions.

- The fourth week requires the implementation of perspective-taking practice.

Find Your Style:
- Adjust your thinking practices to match your personal characteristics.

- Visual learners benefit from *mind maps and diagrams* because these tools help them think more effectively.

- When using social processors, think out loud with colleagues or friends who trust you.

- The structured approach involves using checklists and frameworks.

- The intuitive approach focuses on emotional understanding combined with instinctual assessments.

Reframe the Purpose:
- The main objective of critical thinking is to *understand and improve* instead of seeking fault.

- In order to ask questions, maintain curiosity rather than displaying skepticism.

- Look for ways to build on others' ideas

- Your primary focus should be on solving problems rather than proving your rightness in discussions.

Balance Analysis with Action:
- Perfection in thinking does not excuse the need for making timely decisions.

- The analysis period needs time constraints to be established.

- You must determine when your information collection has reached its maximum point.

- Learn to make choices even when the situation remains uncertain.

Track Your Progress

Weekly Check-ins:

- Which habits did I practice consistently?

- What areas of my thinking improved after the exercise?

- What obstacles did I encounter?

- Which changes should I implement for better results?

Monthly Reviews:

- The level of my thinking enhancement during the previous month needs evaluation.

- What selections did I choose differently through these practices?

- My relationships received what kind of changes?

- The emergence of fresh challenges together with new possibilities became visible.

Quarterly Assessments:

- The assessment areas from your original assessment should

be evaluated for a second time.

- I should recognize my progress and mark new opportunities for development.

- You must adapt your practice methods through acquired knowledge.

- You need to establish targets for the upcoming quarter.

Building a Community of Critical Thinkers
At Work:
- Team meetings should include "devil's advocate" roles for proposal.

- Book clubs that focus on decision-making and critical thinking should begin operation. Book clubs are beginning to become more a part of our culture, and there's even a movement for "quiet reading," which is people in groups in a restaurant or other setting, reading books quietly.

- The organization should establish groups for discussing complex business-related issues.

In Personal Life:
- You should identify companions who enjoy engaging in deep intellectual discussions.

- You should join organizations that combine learning activities with discussion sessions.

- Join online discussions through community forums and local debate societies.

Online Communities:

- Follow respectful thinkers who challenge your current beliefs.

- Participate in forums that utilize evidence-based discussion methods to enhance your learning journey with others while maintaining humility during the process.

- Share your learning journey with others, but maintain humility during the process.

Ask Better Questions:

- Instead of *"Don't you think...?"* you should say, *"What do you think about...?"* This is a wonderful way to keep a conversation going.

- When someone makes a wrong statement, ask for their point of view rather than saying their statement is incorrect.

- Instead of arguing, you should ask for evidence that would change your mind.

Share Your Thinking Process:

- Present the complete logic behind your conclusions along with your findings.

- I should reveal both my doubts about uncertain matters and my changes in perspective.

- People who show genuine interest in matters they don't grasp should be supported.

Create Safe Emotional Spaces for Disagreement:

- The team needs to agree on specific guidelines for maintaining respectful dialogue.

- Before trying to convince someone, it is essential to establish mutual comprehension.

- Search for points that both parties can agree on and fundamental beliefs that unite them.

Maintaining Long-Term Growth
Read Widely:
- You should expose yourself to diverse viewpoints together with multiple academic fields.

- The authors who challenge your current beliefs should be part of your reading selection.

- The content should include a mix of contemporary news and lasting knowledge.

Seek Feedback:
- Seek trustworthy friend evaluations about how your thinking and decision-making processes have evolved.

- Request detailed feedback regarding your questioning methods as well as your communication techniques. People may be a bit reluctant at first, but if they see you're receptive to even negative feedback, they will proceed.

- You should accept criticism that targets your reasoning approach. Yes, this is where humility comes into play.

Teach Others:

- The process of explaining thinking concepts to others strengthens your comprehension of these concepts.

- The practice of mentoring others maintains your commitment to following proper procedures.

- The process of teaching forces you to define the essential value of these skills.

Adapting to Change
- Stay updated about field developments that influence decision-making processes. You want to be ahead of the curve here.

- Acquire knowledge about new tools and technological advancements, which can boost your thinking capabilities.

- An individual should grasp how changes in society influence both information systems and communication systems.

Remain Flexible:
- Your established practices need adaptation whenever your personal life situation transforms.

- When previous methods stop producing results, you should try alternative approaches.

- Be prepared to let go of practices that no longer serve you.

Keep the Big Picture in Mind:
- Critical thinking exists to improve quality of life rather than being an independent goal.

- The focus should be on crucial outcomes, which result in

enhanced relationships along with wiser decisions and increased fulfillment.

- Keep your sense of humility regarding your thinking abilities, but maintain your dedication to improvement.

Chapter 10 Skills Checkpoint

Before continuing, ensure that you can perform the following tasks:

- You should develop particular daily routines and weekly and monthly habits that align with your current lifestyle.

- Learn methods to handle typical challenges, including time limitations, memory failures, and resistance to change.

- Construct supportive settings that enable positive thinking routines.

- A proper tracking system should be implemented to monitor improvements and adjust practices based on results.

- The integration of thinking skills requires consistent application of critical thinking habits across all life areas.

- **Self-Check Questions**

1. Which essential critical thinking practices would create the most beneficial effect on your everyday life?

2. What challenges do you expect to face when practicing these habits, and what strategies will you use to overcome them?

3. Which members of your social network would help you develop critical thinking abilities, and what methods would you use to help them achieve theirs?

Real-World Application Challenge:

Create a personal critical thinking system that you will execute:

1. Assess your current critical thinking abilities across various life domains.

2. Select 1-2 critical areas that demand your immediate attention for improvement.

3. Select 3-4 particular practices for your implementation.

4. Establish cues together with reminders, which will enhance your new habits.

5. You need to select at least one person who will support your development journey.

6. Create a simple system for monitoring progress

7. *Make a practice commitment* to continue for thirty days with scheduled weekly meetings.

Start with modest steps while maintaining consistency through self-adjustments based on your personal discoveries.

Critical thinking exists as an ongoing, lifelong process rather than a target destination that you achieve. Just as Rachel learned critical thinking needs constant practice. You can develop habits that maintain your thinking acuity and improve your judgment even when external factors transform.

How do you develop a habit? Habits come almost automatically once you begin to use them repeatedly. I remember learning this when I was taking basic self-defense classes in karate. We learned how to raise one arm to defend our faces, and two decades later, I still have that action if anything comes toward me, and I think I might be hit by it. I don't even have to think; it just happens.

The main objective is to advance toward better thinking abilities and decision-making and communication skills through daily improvements. You might think of it almost as **learning a new language**. At first, the words come haltingly, and that's what may happen with critical thinking, but after you use it, it becomes much easier, and you become fluent. It's even like learning to read when we first went to school. Remember having to learn the alphabet, how to pronounce the letters, and then to put them into words? It took time, and we had to keep ourselves motivated by the books that were given to us.

Chapter 11:
The Power
of Perspective:
Reframing as a
Critical Thinking
Tool

Reframing represents one of the most powerful tools in the critical thinker's arsenal, fundamentally altering how we perceive and interpret information, problems, and situations. At its core, *reframing involves deliberately shifting our perspective or changing the conceptual framework through which we view a particular issue.* This cognitive strategy allows us to break free from limiting assumptions, discover new possibilities, and often identify solutions that were previously invisible to us.

The psychological foundation of reframing rests on the understanding that our initial interpretation of events is not necessarily the most accurate or useful. Human cognition naturally seeks patterns and meaning, but these mental shortcuts can sometimes *trap us in narrow thinking*. When we reframe, we consciously step outside our automatic interpretive patterns and consider alternative ways of understanding the same information. This process can transform what appears to be an insurmountable obstacle into a manageable challenge or reveal opportunities hidden within apparent setbacks. But it means consciously doing this.

Effective reframing often involves questioning the assumptions embedded in our initial understanding of a situation. For instance, *what we initially perceive as a personal failure might be reframed as a valuable learning experience* or necessary preparation for future success. Similarly, a constraint or limitation might be *reframed as a creative challenge* that sparks innovation. The key lies in recognizing that *multiple valid interpretations* of the same set of facts can coexist, and our choice of frame significantly influences our emotional response and subsequent actions.

The practical applications of reframing *extend across numerous domains of human experience.* In problem-solving contexts, reframing can help us move beyond obvious but ineffective solutions by encouraging us to reconsider what the real problem actually is. Sometimes what appears to be a technical challenge is actually a communication issue, or what seems like a resource problem is really a priorities problem. In interpersonal conflicts, reframing can help us shift from adversarial thinking to collaborative problem-solving by changing *how we understand the other party's motivations and actions.*

Critical thinkers must also remain aware of the potential pitfalls of reframing. While this cognitive tool *can be liberating and productive, it can also be misused to avoid uncomfortable truths or justify poor decisions.* **The key is to exercise caution**. The goal is not to find the most pleasant interpretation of events but rather to find more accurate, useful, or comprehensive ways of understanding them. Effective reframing requires intellectual honesty and a commitment to evidence-based reasoning rather than wishful thinking.

The skill of reframing develops through practice and conscious attention to our own thinking patterns. By regularly asking questions such as *"What other ways could I look at this?"* or *"What assumptions am I making that might not be true?"* This way, we can cultivate greater cognitive flexibility. This mental agility becomes particularly valuable in our rapidly changing world, where the ability to **adapt our thinking** to new circumstances often determines our success in both personal and professional endeavors. Ultimately, reframing empowers us to become active architects of our understanding rather than passive recipients of first impressions.

Managing Emotions Through Strategic Reframing

When engaged in challenging discussions or debates, our emotional reactions can quickly derail our critical thinking abilities (*remember that **System 1** we mentioned earlier?*). Learning to reframe situations in real-time helps maintain cognitive clarity and *prevents defensive responses* that undermine productive dialogue.

Consider the moment when someone directly challenges your cherished belief or questions your expertise. Your initial frame might be, *"This person is attacking me and disrespecting my knowledge."* This interpretation naturally triggers defensiveness and anger. However, you can immediately reframe the situation as, *"This person is giving me an opportunity to test and strengthen my reasoning"* or *"They're helping me identify potential weaknesses in my argument that I should address."*

This shift transforms a perceived threat into a valuable intellectual exercise.

I will never forget the first boss that I had right out of school. He was one of the mildest-mannered, unflappable people I have ever encountered. He would listen, think, and then quietly provide either a question or an answer to something. You always left his company thinking you were better for what he had done.

Occasionally, I wonder where he developed that skill, and in reading some material about him, I found the possible answer. He had been in World War II as a member of the Army Air Corps. Having witnessed death and destruction, the man was able to reframe his perspective on life.

He never made anyone look foolish and always complimented them for what they were including in any discussion. When someone presents evidence that contradicts your position, resist the initial frame of "*They're trying to prove me wrong and make me look foolish.*" Instead, reframe it as "*I'm gaining access to information I didn't have before*" or "*This is a chance to update my understanding with better data.*"

This approach keeps you focused on learning *rather than defending your ego*. It did, in fact, help me to develop my personal theory of what I call "*the crumb theory.*" What is that? Any meeting or discussion I've attended is only going to provide me with a tiny bit of new information. I never go in expecting an incredible presentation or something that will turn the world around. All I want is a little crumb, and from there I build. It's never failed me. In fact, I wrote a blog on Medium.com on it, and if you do a search on the internet, you'll probably find it. It's especially useful when you are writing or wanting to learn something.

My expectation of only seeking a crumb of new information or new learning has proven to be invaluable to me. I went into that seminar expecting to hear a required presentation on child abuse and I came away with something that led me to write a professional article that then resulted in my appearing on national TV to discuss it. In effect, it led to my being a sought-after television. expert on anxiety and stress disorders for the next decade. That, in turn, earned me a book contract and authorship of an important chapter in a clinical psychology handbook. I never thought that a seminar I had to attend would mean a major career push for me. The crumb I came away with was well worth the two hours I spent at that presentation.

If you notice yourself becoming frustrated when others seem to miss your points, step back from the frame *"These people just don't understand"* and consider *"I need to find a clearer way to communicate this idea"* or *"Their questions are revealing gaps in my explanation."* This reframe shifts responsibility back to you as a communicator rather than blaming your audience.

A university professor informed a friend of mine, who was taking a test that required writing a lengthy essay, about the importance of clarity in communication. The professor looked at the answer, then told the student, *"You know what you're writing about, and you know it thoroughly, but your reader doesn't. **Make it clear to the reader."*** That's another simple rule to keep in mind.

When facing aggressive or dismissive behavior from others, avoid the natural frame *"I need to put this person in their place."* Instead, try *"This person's emotional reaction might indicate they feel threatened by the topic"* or *"Their behavior is information about their mindset, not a reflection of my argument's merit."* This reframe helps you respond strategically rather than reactively. Keep your emotions in check.

Practice reframing interruptions or tangents not as disruptions, but as opportunities to understand how others process information differently. When someone brings up seemingly irrelevant points, reframe from *"They're derailing the conversation"* to *"They're showing me connections I hadn't considered"* or *"This reveals their underlying concerns that we should address."*

The key is developing quick mental habits that automatically search for alternative interpretations when you feel your emotions rising. This emotional regulation through reframing keeps your critical thinking faculties engaged when you need them most.

Chapter 12: Using Critical Thinking to Master Job Interview Success

Josephine, walking into her dream job interview at a growing tech company, was not just another hopeful candidate. She'd become a critical thinker—someone who questions, analyzes, and connects the dots that others miss. What's more, she knew the CEO had started his career in education before pivoting to technology because she asked herself, "*What drives this leader's decisions*?"

Understanding that the company had just secured major funding to expand into healthcare software was one more thing she knew because she evaluated multiple news sources and identified the pattern. Preparation provided her with the knowledge that she knew about the recent industry report predicting 40% growth in their market sector because she synthesized information from different sources to see the bigger picture. She did her homework.

Josephine didn't stumble upon this information by luck—*she developed it through critical thinking*. Having learned to question assumptions, analyze evidence, and draw logical conclusions, she now had the answer in her grasp. And what did she discover? She got the job.

Most people think preparing for an interview means memorizing answers to common questions. *That's surface-level thinking*. Critical thinkers know that success comes from *developing a deep understanding of the situation*, evaluating information from multiple angles, and making connections that reveal insights others miss.

I know that when I prepared for interviews to write magazine articles, I diligently reviewed all the files we had on that company and the individual involved. When I went for the interview, I can remember people being astonished at what I knew about their company. One woman couldn't believe I knew she had a pony when she was 12 years old.

Here's what research tells us: 47% of hiring managers will not offer a job to someone who doesn't demonstrate knowledge about the company. But here's what critical thinking reveals: when you analyze information systematically and draw meaningful conclusions, you join the 20% of candidates who truly stand out.

Critical thinking isn't just a fancy term—it's a practical skill that *transforms how you approach any challenge*, including job interviews. It's about asking the right questions, evaluating information critically, and synthesizing what you learn into actionable insights. Let's explore how to apply these skills to master interview preparation.

Critical Analysis: Questioning What Companies Really Tell You

Think of company research as an exercise in critical analysis—the skill of breaking down complex information into manageable parts and examining each piece carefully. Most job seekers accept company information at face value, but critical thinkers dig deeper.

Start with the obvious sources, but don't stop there. The best way to research a company is to *review its website* and search for *news articles or blog posts* about it. But here's where critical thinking kicks in: *treat every piece of information as a claim that needs verification.* Ask yourself, "*What is this source trying to convince me of, and what might they be leaving out*?" What's left out might, in fact, be the most important thing you can gather.

The company website is like a carefully constructed argument designed to persuade you of its excellence. Critical thinkers recognize this bias and seek multiple perspectives. They ask probing questions: What challenges is this company facing that they're not highlighting? How do their stated values translate into actual practices? What evidence supports their claims about innovation or growth? Question, question, question.

This is where your analytical skills become powerful. Look at their mission statement, but then practice synthesis—combining information from different sources to form a complete picture. If they state that they value work-life balance, how do employee reviews on *Glassdoor* reflect this? If they emphasize innovation, *what patents have they filed recently*? What *innovative projects* can you find evidence of in *trade publications*? Never overlook the trade press.

Most trade publications have a website, but one of the great resources that we have in our communities is the local library. The reference librarian can be invaluable in pointing you not only to trade press publications, but to indexes that carry information on particular industries. I never knew about indexes until I asked a research librarian, and she opened my eyes to their benefits.

Critical thinkers also know how to identify credible sources. When you search recent news coverage and read press releases, evaluate the source's reliability and potential bias. Is this information coming from the company itself, independent journalists, or industry analysts? Each source has different motivations and different levels of objectivity. you might even ask, *"Who owns that particular publication?"* and *"What might their orientation to business be currently*?"

Here's a critical thinking exercise that most job seekers skip: *analyze their competitors*. This isn't just helpful—it's essential for developing what critical thinkers call "context." You can't truly understand a company's position without *understanding the landscape it operates in*. Ask yourself: What makes this company different from its competitors? What challenges do all companies in this industry face? What opportunities exist that this company is positioned to capitalize on?

The most sophisticated critical thinkers also practice what researchers call *"perspective-taking."* Don't just look at what the company says about itself—investigate what others say about it. Check employee review sites, but read them critically. What patterns emerge when you analyze multiple reviews? Are complaints isolated incidents or systemic issues? How does the company respond to criticism? Has anyone come forward with information? Here we might even look at whistleblower reports and what type of attention they have received and the legal results of any of that.

Recall that the goal of critical thinking is to comprehend reality so that you can make wise decisions, not to identify flaws.

Assessing Sources: Examining Professional Profiles in a Different Way

Now, let's use critical evaluation techniques to investigate potential coworkers and superiors. Understanding the professional environment, or the culture you want to join, and the traits that contribute to success there is more important than passing judgment on others.

A remarkable culture of camaraderie existed at Bell Telephone Labs, which produced some of the most significant advances in computers (and transistors). Important discoveries emerged from the casual conversations that engineers had with their peers as they strolled through the rows of cubicles. The company encouraged its engineers to have casual conversations at any time during the day and offered a large, free cafeteria for all of them to eat together. I know much of this information because it was one of the first jobs that I got out of school, and I saw it all first hand.

You can practice evaluating sources and identifying patterns on LinkedIn. Investigate the hiring manager's background on LinkedIn, but approach their profile as you would a critical thinker evaluating data. What can you learn about the company's values from this individual's career path? Which abilities and backgrounds appear to be valued here?

In the LinkedIn search bar, type the name of the business and keywords associated with the hiring manager's position. Choose 'People' to filter the results, then 'Current Company' to select the desired company. Here's where critical thinking turns ordinary research into insightful knowledge, though: look for trends in several profiles.

Apply critical thinking to your findings. It appears that the company promotes from within if you notice that the majority of senior leaders began their careers in entry-level roles. Executives' hiring preferences become apparent when their educational backgrounds are similar. A culture that values thought leadership is indicated by the fact that leaders regularly post about industry trends.

Examine the subtleties to improve your inference abilities. What does it indicate about the organization's professional development policies when you notice that a hiring manager has switched between departments? What can you conclude about the prospects for advancement when you observe someone has received a rapid promotion? Are these managers also on community boards, such as education and development?

The most insightful information is obtained through synthesis, which is the process of integrating data about several individuals to comprehend the overall picture. What kind of schooling do the executives possess? You may notice a pattern in the kinds of people that businesses prefer to hire. This pattern recognition aids in understanding not only the skills they seek but also the type of thinking that is appropriate for their culture. One university that I am aware of **never employs its graduates**. For what reason? They are looking for a wide variety of viewpoints, and adding people from other universities broadens their current pool.

According to research, your chances of getting the interview are doubled if you mention a shared set of values with the organization. Critical thinkers understand that the goal of this process is to find real connections based on analysis and evidence, not to manipulate.

Strategic Forecasting: Using Evidence to Predict Industry Futures

The ability to assess forecasts, analyze trends, and form well-informed opinions about potential futures is where critical thinking really shines. Anyone can learn the history of a company by heart, but critical thinkers know how to predict the direction of industries and how that affects career choices.

Job market analysis looks at workforce dynamics, industry growth trends, and employment trends to give important information about the state of the labor market today. It is your responsibility as a critical thinker to assess this data, spot trustworthy trends, and make reasoned judgments about potential future opportunities. Will this company expand, or is it on the decline?

Because government resources like the U.S. Bureau of Labor Statistics employ peer review and systematic methodology, they are especially valuable. The **Occupational Outlook Handbook** is one of the additional resources that the government offers. You're getting analysis based on rigorous research methods when they release employment projections that examine trends in construction job growth brought on by the expansion of renewable energy or the need for AI data centers.

As critical thinkers say, practice "trend analysis." Is the industry expanding or contracting, and if so, what particular factors are causing this shift? What new skills are becoming important, and what evidence supports this conclusion? What challenges is the industry facing, and how might those create opportunities for people who think strategically?

Apply cause-and-effect reasoning to analyze demographic and economic trends. An aging population doesn't just create opportunities in healthcare—it also drives demand for financial planning, home modification services, and transportation alternatives. Critical thinkers make connections between these various factors and identify opportunities that others may overlook.

The most sophisticated critical thinking skill is learning to distinguish between reliable predictions and wishful thinking. Just because someone predicts dramatic growth doesn't make it likely. Evaluate the assumptions behind predictions. Are they based on historical patterns, current evidence, or hopeful speculation?

Here's a critical thinking exercise: take any industry trend you've identified and trace it backward. What caused this trend? How long has it been developing? What obstacles might prevent it from continuing? This kind of analysis helps you evaluate whether a trend represents a genuine opportunity or just temporary hype.

Do you know which simple graph *Alan Greenspan* (Chairman of the Board of Governors of the Federal Reserve System) used to make his economic predictions? And not everyone knows, but he completed his doctorate at NYU only after he had become a success in business and consulting. What was his secret to success? *He analyzed the sales of wallboard and nails* used in construction. From there, he could extrapolate the growth of the housing/building construction industry, which meant job opportunities and corporate profits in that area and it expanded outward into all the industries needed to provide that increase in construction. . While everyone else was looking at major indicators, Greenspan looked at the minutiae, and that's where he found the gold. What minutiae might lie in the corporate statements where you are applying for a job?

A successful investor I knew always analyzed natural catastrophes, for example Hurricane Katrina, to identify industries that would experience a surge in demand for their services or products. It was always a *"good from bad"* scenario. He knew they would have to build roads, infrastructure, drainage, cell towers, and much more, and he invested in all of it. Using this type of thinking, he managed not only to support his mother and three sisters but to put himself through medical school.

I was enrolled in a college evening oceanography course, and I had no idea that the professor would give me important career advice. No, it was about careers in general rather than oceanography. He looked at us, a world-renowned expert on a tiny kind of snail, and said, *"Find your snail."* For us, what did that mean? He reassured us that discovering what others overlooked or thought was unimportant would shape our futures. He suggested to us that we *"develop that area into one where you are the ultimate expert,"* I have always searched for my snail, and I will never forget. Didn't Alan Greenspan do that?

Combining Knowledge: Using Research to Gain a Competitive Edge

The value of all this critical thinking depends on your ability to combine your findings and use them wisely. The objective is to show that you can think critically, assess information critically, and come to logical conclusions, not to impress others with the amount of research you have done.

Make connections between disparate pieces of information to show off your critical thinking skills during a job interview. Don't just share what you've learned when they ask **why you want to work there**; instead, describe the reasoning behind your decision. Demonstrate how you assessed industry trends, analyzed their competitive position, and concluded that this company provides the best opportunity for a candidate with your goals and skill set.

I knew a young man who was applying to go to graduate school and he was sent for an interview with a member of the board of directors of the university. Sitting down, he quickly discovered that all the board member wanted to talk about was the university and his personal experiences. The young man had no experience with that university and knew very little about it other than it had a fine program in which he was interested. He went away wondering what he should have done. In fact, the board member was trying to figure out if this young man had the same values as he about the university. He was not offered admission.

Asking insightful questions that elicit deeper understanding is another skill that critical thinkers possess. Rather than posing general inquiries such as *"What is your company's mission?" "Given the industry trends toward automation, how is your company positioning itself to maintain its competitive advantage?"* is an example of an analytical question. This demonstrates your strategic consideration of their opportunities and challenges.

Keep in mind that intellectual humility—realizing your knowledge gaps and remaining receptive to new information—is another aspect of critical thinking. Employers appreciate the way you think when you demonstrate that you have done extensive research while still being interested and ready to learn more.

The confidence derived from wishful thinking is not the same as the confidence derived from systematic analysis. That conviction comes through when you've given a company and industry a critical evaluation and determined that this is the place where you want to develop your career. You're demonstrating once more that you possess the critical thinking abilities necessary to make wise career choices, not merely hoping they'll pick you.

Although only 24% of applicants are satisfied with the interview process, the entire process is changed when preparation is viewed as a critical thinking exercise. Not only are you responding to their inquiries, but you are also exhibiting the critical thinking abilities that are necessary for success in any position. It's a no-brainer.

You can cultivate the analytical abilities that set exceptional candidates apart from mediocre ones, just like Josephine did when she approached her career opportunity with a critical mindset. You can examine the data. The patterns are just waiting for you to notice them. You just need to make the connections that are already there. Chapter 14 is going to help you make an additional tool available to yourself, and that is mind maps.

The question isn't whether you can find information about companies and industries—anyone with internet access can do that. The question is whether you can think critically about what that information means and use those insights to make better decisions about the future. Also a question of knowing how the interview will progress and what methodology (Janz) they use as their interview framework.

Your future self will thank you for developing these thinking skills today. Begin by asking questions, engaging in analysis, and making connections that others may overlook. Most importantly, start thinking like the critical thinker you're capable of becoming.

Chapter 13: Your Next Steps

M aria became overwhelmed by the vast amount of conflicting advice, along with the many decisions and flood of information that dominated her daily activities after she first touched the book. Better thinking skills were essential to her, yet she remained unsure about her starting point and doubted her ability to transform her mental patterns after building them through decades.

Through nine chapters, Maria discovered that critical thinking represents functional mental tools that simplify life by creating a clearer understanding, easier choices, and stronger relationships. She has learned to delay accepting information when it supports her preferred beliefs. She asks questions that generate more useful discussions. And there was more because she could handle important choices with improved self-assurance, although she couldn't forecast all outcomes. What was she able to do? For one thing, she maintains her mental acuity through persistent thinking practices, avoiding constant questioning and doubt.

The main aim of this book is illustrated through Maria's case, as the combination of specific skills and consistent practice leads to significant improvements in thinking abilities, decision-making, and communication skills. The tools you've learned are not academic concepts because they function as useful practical methods that *help you in everyday situations*, from family gatherings to workplace issues and important life choices.

What You've Accomplished

Foundation Skills You've Built:

• Simply by reading this book and doing the exercises faithfully, or even once in a while, you have built a storehouse on which you can rely. You have gained knowledge about your thinking process, the automatic mental patterns that mislead you, and when to evaluate purposefully.

• Even when you think that there is too much going on and there is just too much information to handle, you now have a way to separate out the important and the less important. You have gained new skills, gained a means to evaluate more efficiently, and developed the ability to cultivate even better habits in your skill set.

• You learned to distinguish between information availability and knowledge of reliable, relevant information.

Core Thinking Skills You've Developed:

You now ask questions that both uncover hidden assumptions and explore alternative perspectives to gain more profound insight instead of verifying your existing beliefs. Now you aren't a victim of confirmation bias.

You have learned to evaluate evidence through systematic processes that help you identify strong research versus convincing claims while understanding which conclusions you can and cannot draw from various types of information.

You identify logical fallacies in arguments while creating constructive responses instead of simply focusing on pointing out errors.

Advanced Application Skills You've Mastered:

- You have learned to make sound decisions under uncertain conditions by using structured frameworks together with scenario evaluation and outcome-based learning instead of waiting for flawless information.

- You combine your ability to clearly communicate complex concepts with your ability to reach consensus through dialogue rather than debates.

- You have learned to work effectively with AI tools as thinking partners while preserving your own independent judgment along with human wisdom.

Life Integration Skills You've Established:

- Your daily practices provide lasting benefits to your critical thinking abilities while requiring minimal time commitment and disrupting no part of your daily life.

- You have established conditions in your surroundings and personal connections to promote your thinking abilities and your decision-making skills.

- Your critical thinking method adjusts to various environments while keeping essential principles intact.

The Compound Effect of Better Thinking

The true power of critical thinking skills develops over time because of what we can term the "compound effect" of better thinking. The process of compound interest helps money grow, while small enhancements in your thinking abilities and decision-making and communication skills result in larger cumulative benefits throughout multiple months and years.

Career and Professional Growth:

Every organization benefits from employees who demonstrate better critical thinking abilities. Insightful questioning, decision-making, persuasive communication, and problem-solving skills make you more valuable. Your colleagues now look to you for your input. Managers will have confidence in your ability to make sound decisions. Coworkers and executives will detect your ability to analyze complicated matters, which leads to new opportunities. People may come to you for help with their decision making process, and you will have the tools.

Financial Well-being:

The development of critical thinking abilities enables you to **prevent costly errors** while helping you discover authentic opportunities. Improved financial thinking skills help you resist scams, make better investment choices, avoid unwanted purchases that contradict your values, and enable more effective planning for future goals. Better financial thinking allows you to save and earn money, which builds up substantially over time. Retirement is always an issue, and you will have prepared for it well into your past for your future.

Relationship Quality:

Your ability to ask better questions, listen more thoughtfully and communicate better will lead to **better relationships**. Less confusion exists, and conflicts end in positive ways while strengthening your relationships with significant others. The enhanced bonds you establish through stronger relationships build support networks that enhance all aspects of your existence. And we know that support systems are vital to our physical and mental health.

Personal Satisfaction:

The ability to think critically enables you to make choices that correspond to your core values instead of letting external forces or instinctual responses determine your actions. Devoting your time and energy toward meaningful things gives you satisfaction and prevents unnecessary remorse.

Common Challenges and How to Navigate Them

"I Feel Like I'm Overthinking Everything."

Analysis should help you make better decisions instead of creating more challenges. The thinking approach should differ between situations, since you need quick and intuitive choices for minor decisions and more thorough evaluation for crucial ones. Time constraints should be established for decision-making processes because "good enough" solutions usually surpass perfect outcomes.

"People Think I'm Being Difficult When I Ask Questions."

When you maintain an appropriate tone and genuine intentions, others will not perceive your questions as difficult or argumentative. Ask questions out of genuine interest instead of showing skepticism. Replace statements starting with *"Don't you think"* with *"Help me understand..."* and *"I'm wondering whether..."* to communicate effectively. The ultimate goal should be to understand better instead of winning disputes. The skill of *"Help me understand"* is something that therapists have used for decades; it opens up communication that would have been closed.

"I Keep Falling Back Into Old Habits."

The process of changing habits proves difficult as well as being unpredictable. *Treat all setbacks as learning experiences* because they provide valuable lessons instead of considering them as failures. Begin with minor alterations, then create environmental signals that promote new behaviors while acknowledging progress instead of seeking flawless results. The key factor for success lies in maintaining consistent effort over time instead of achieving flawless daily performance.

"I Don't Have Time for All These Practices."

You should not feel required to implement all techniques mentioned in this book. Select 2-3 techniques that target your most significant problems and implement them within your current daily activities. Critical thinking enhancements mostly require mindset transformations instead of requiring extra time dedication. A brief thirty-second pause during information sharing or decision-making produces substantial effects.

"My Family/Workplace Doesn't Support This Kind of Thinking."

Begin by demonstrating better thinking behaviors instead of attempting to transform others. Focus your efforts on developing better decision-making abilities while asking improved questions. Your better thinking results will typically generate curiosity from those around you who wish to learn your methods. Creating a supportive community starts with finding just one person interested in clearer thinking.

Advanced Resources for Continued Learning (All sources are listed in the reference list at the end.)

Books That Build on These Foundations:

• For deeper exploration of decision-making: "Decisive" by Chip Heath and Dan Heath, "Thinking in Bets" by Annie Duke

• For advanced logical reasoning: "Being Logical" by D.Q. McInerny, "The Art of Reasoning" by David Kelley

• For communication and persuasion: "Made to Stick" by Chip Heath and Dan Heath, "Getting to Yes" by Roger Fisher and William Ury

• For bias awareness and behavioral economics: "Predictably Irrational" by Dan Ariely, "The Righteous Mind" by Jonathan Haidt

Online Communities and Resources:

• Locate discussion groups or book clubs or debate societies which emphasize thoughtful conversations in your area

• Seek writers who present challenges to your beliefs while maintaining intellectual humility

• Take part in online forums that focus on evidence-based discussions and rational thinking

• Take online courses in logic and statistics and decision science through established academic institutions

Professional Development Opportunities:

• Many organizations provide their employees with training programs that focus on decision-making skills, problem-solving methods and communication techniques

• Most professional organizations organize workshops about analytical thinking that align with your professional field

• The pursuit of additional education in statistics and philosophy and debate and scientific method can be valuable for those who want to develop their critical thinking abilities

Staying Current in a Changing World

The fundamental principles of critical thinking will persist with specific technologies and information sources, along with challenges. Instead of learning specific techniques by heart, the emphasis should be on *developing flexible skills*.

Technology Changes:

The sophistication of AI tools will increase, yet your ability to ask good questions, evaluate evidence and preserve independent judgments will gain importance.

You should *maintain an interest in new technology* while keeping a realistic view of its potential limitations. Have you subscribed to any newsletters that will keep you up-to-date on artificial intelligence? This type of communication is an area that is of prime importance. In fact, this area is moving forward at a faster pace than anyone could have expected. It may well become the career choice of people who wish to advance quickly. We have already seen that simply learning computer programming is insufficient in an AI-generated business environment.

Information Environment Changes:

The ability to analyze sources along with evidence and critical thinking skills about claims will remain effective regardless of the communication platform.

Social and Political Changes:

The society will experience different issues, but your ability to explore diverse viewpoints, challenge beliefs, and take part in beneficial dialogues will persist.

The Ripple Effect: Thinking Better Together

Critical thinking development brings the most rewarding experience by showing how your enhanced mental abilities influence others. Your thoughtful inquiry triggers deeper thinking in everyone nearby. Intellectual humility, which you demonstrate, will make others feel comfortable acknowledging their uncertainty while changing their perspectives. Effective communication combined with attentive listening produces a more productive dialogue.

The practice of critical thinking does not require you to advocate for others to adopt your methods or to correct their thinking processes. The act of developing these skills personally will generate beneficial effects that enhance better thinking across your household, work environment and neighborhood.

In Your Family:

Children learn more by watching than by being directly instructed. Observing your deliberate thinking pauses, your practice of questioning before judgment, and your willingness to admit uncertainty allows your children to learn these behaviors naturally. Parents and adults are children's first teachers.

In Your Workplace:

Any team that includes a single person who maintains consistent inquiry into facts, communicates effectively, and considers multiple viewpoints will perform at higher levels. Your established thinking patterns have the power to change the way meetings function and influence the decisions made and problem-solving methods adopted by your organization.

In Your Community:

The active involvement of citizens who think critically and seek reliable information leads to improved civic discourse, which results in better community solutions.

Your Personal Critical Thinking Manifesto

As you progress in developing these skills, it would be beneficial to develop personal rules for your thinking process and decision-making approach. You can modify this outline to create your personal critical thinking guide.

I commit to:

• I will not rush to accept information that perfectly matches my existing beliefs.

• I ask questions because I am curious rather than to validate my arguments.

• I seek different viewpoints on matters important to me.

• I recognize my knowledge gaps and acknowledge instances when my opinions could be incorrect.

• I choose to decide through evidence and my values instead of letting emotions and social forces determine my choices.

• I present my thoughts in simple language while actively listening to different opinions.

• I will learn new information and update my understanding through my experiences.

I will remember that.

• It is more important to grasp the truth than to claim rightness.

• Following beneficial evidence to change your mind shows intellectual strength rather than weakness.

• Effective questions typically have more value than the speed at which answers are delivered.

• The feeling of not knowing is normal in our complicated world.

• My thought process affects not only my personal life, but it also shapes the reality of those who surround me.

The Journey Continues

Your critical thinking development continues beyond this book's conclusion. The abilities you have learned function as instruments to support your ongoing development of improved thinking abilities, improved decision-making, and enhanced communication throughout your lifetime.

Occasionally, you will notice yourself slipping back into previous behaviors by accepting information too fast or making choices based on emotions or not expressing yourself clearly enough. Such thinking is normal and expected. The primary task is to detect these instances and use positive techniques to steer your thoughts toward better methods.

On other days, you'll think with remarkable clarity, asking questions that unlock new understanding or deciding with confidence despite uncertainty. These experiences help you understand the significance of your growth journey and drive you toward additional development.

Maria's journey started with feeling overwhelmed by information while doubting her decision-making abilities. She developed self-assurance in her ability to handle complex topics by practicing the book's skills regularly.

Your story will differ since your obstacles, along with your situation and objectives, are distinct. The core guarantee remains unchanged because learning and applying critical thinking skills enables you to navigate the complex world with greater wisdom, along with increased confidence and effectiveness.

These thinking tools from the book will become daily gifts that give you a clearer understanding, better decision-making abilities, and more meaningful relationships with others. Use these skills properly while distributing them freely to maintain your growth as a thinker from this day forward.

Your path as a critical thinker is developing now. Stay on the path; return whenever you need a refresher, and enjoy your wonderful journey.

Chapter 14: Mind Maps and Critical Thinking

Have you ever tried to solve a problem by writing down everything you know about it on paper? Most people just make lists or write paragraphs. But there's a **better way that can actually make you think more clearly and solve problems faster.** It's called **mind mapping**, and it's one of the most powerful tools you can use to strengthen your critical thinking skills.

Why are mind maps so useful? For one thing, not all of us benefit from strictly written words. **Some of us, myself for instance, are visual thinkers** and we see things visually in diagrams that may have an impact. In fact, I have an unusual ability, I call it a gift. When I can't remember something, like someone's name, guess what I can remember—yes, I can remember the number of letters in their name. That sounds a bit bizarre, I know, but I'm going to reframe it as a "gift" and not something to turn your nose up at.

I also can manipulate visuals mentally and turn them around and see them at different angles. That, I believe, is a very unusual gift, and I love it. This latter gift has enabled me to design things in my mind, put them together and make a whole of them. I suppose that's what people do who are working on patents for inventions they've developed. I'm not an inventor, but I am a highly visual thinker. For me, mind maps are a gift, and I love every single one of them because they help me to quickly see connections. I hope it's the same way for you.

What Are Mind Maps?

A mind map is like a picture of your thoughts. Instead of writing ideas in straight lines like a regular list, you put your main idea in the center of a page and draw branches out from it. Each branch represents a different part of your main idea. Then, you can add smaller branches to show how different ideas connect to each other.

Think of it like a tree. Your main topic is the trunk, your main ideas are the big branches, and the specifics are the smaller branches. This format mimics the natural way your brain functions, so it's not just a fancy way to take notes. The way your brain works isn't linear. It jumps from one idea to another, connects concepts, and notices patterns everywhere. Algorithm does the same thing; notice patterns and they are built on how our brain functions and they mimic our brain. .

This visual method called mind mapping uses non-linear associations and synaptic connections to *mimic how humans actually think.* You're working with your brain, not against it, when you make a mind map.

Why Mind Maps Encourage Critical Analysis

Critical thinking entails carefully examining information, challenging what you hear, and coming to well-informed conclusions. You can improve these tasks in a number of ways by using mind maps.

First of all, mind maps make you **deconstruct large, complex issues into smaller ones**. It's simpler to identify what's missing, what doesn't make sense, or what requires further research when all the components of an issue are presented graphically.

Second, mind maps make connections that you might otherwise overlook. Students can develop a deeper comprehension of the subject by visualizing the connections between ideas through mind mapping. Drawing connections between related concepts reveals relationships and patterns that may *provide fresh perspectives.*

Third, mind maps push you to think critically about everything. You naturally wonder, *"What else connects to this?"* as you add each branch to your map. *What have I been missing? What connection does this have to that other concept?*

This is supported by research. According to studies, kids in mind mapping experimental groups scored noticeably higher on critical thinking tests than kids in control groups. According to a different study, mind mapping is an effective active learning technique for encouraging critical thinking.

Using Mind Maps to Make Business Decisions

Every day, astute business executives use mind maps to improve their decision-making. Mind maps assist in organizing all the variables involved when a business must make a difficult decision, such as whether to introduce a new product or enter a new market.

To develop a thorough picture of the situation, business analysts employ six strategic categories: ***strategy, service, process, applications, information, and infrastructure***. Consider a small restaurant owner who is attempting to determine whether to provide delivery services. In her mind map, she might center "Delivery Service Decision" and then make branches for:

Costs include delivery trucks, insurance, driver compensation, and packaging.

Benefits include increased sales, a competitive edge, and more clients.

Risks include poor food quality, delayed deliveries, and increased stress.

Resources: marketing budget, employee time, and technological needs

Visually mapping out all of these factors allows her to see the entire picture at once. She might notice that the "technology needs" branch connects "costs" (paying for an app) and "benefits" (making ordering easier for customers). This kind of connection-making is the essence of critical thinking.

Complex deliberation processes are shielded from cognitive confusion by decision mapping. Clear explanations make it much more difficult to miss crucial details or become overwhelmed by information.

Social Science Mind Maps

Mind maps are used by social science researchers and students to comprehend social issues and complex human behaviors. These topics are ideal for mind mapping because they cover a wide range of interrelated topics. Take a look at some of the *mind map resources we've noted later in this chapter.*

Imagine a student researching the reasons behind the disparity in crime rates between neighborhoods. A conventional method might entail taking notes and reading a large number of research papers. However, the student can use mind mapping to place "Neighborhood Crime Rates" in the middle and then branch out to investigate:

- Economic factors include income inequality, poverty, and unemployment.

- *Social factors include family structures, education levels, and ties to the community.*

- *Environmental aspects include public areas, lighting, and housing quality.*

- *Policy elements include zoning regulations, social programs, and police presence.*

Students can better understand the connections between different concepts and spot patterns in the subject matter with the aid of relational and structural mind maps. Patterns begin to appear as the student adds details to each branch. Perhaps crime is higher in places with inadequate lighting and fewer community programs. Perhaps, regardless of income level, crime is lower in areas with good schools and employment opportunities.

Instead of just teaching students facts, this visual method encourages them to think critically about complicated social issues. When using mind mapping techniques, students showed enhanced conceptual understanding and the capacity to formulate well-informed arguments.

Mind Maps in Teaching

Mind maps are used by educators at all levels to assist students in strengthening their critical thinking abilities. Mind maps' visual format accommodates a variety of learning preferences and simplifies difficult subjects.

Mind mapping can be used in a variety of subjects. Students are able to diagram intricate *scientific procedures or experiments*. They can investigate themes, characters, and plot structures in *literature*. They can study the causes and consequences of historical events in *social studies*.

For example, a history teacher in high school might assign students to make mind maps about World War II. Students would place "World

War II" in the center and make branches for the following instead of merely learning dates and battles:

- *Causes include failed treaties, political unrest, and economic issues.*

- *Significant occurrences: pivotal conflicts, pivotal moments, and crucial choices*

- *Repercussions include social, political, and economic shifts.*

- *Key players include important civilians, generals, and leaders.*

Have you ever sat in a history class and all you did was remember the dates of major battles? Have I learned anything by knowing that there was a war in the United States in 1812? What did you really learn? With mind maps, students learn how economic issues caused political tensions, how some battles became pivotal moments, and how decisions made during the war influenced the postwar world as they conduct research and complete their mind maps. This activity promotes independent, critical thinking in addition to reinforcing the material discussed in class.

Mind Maps for Individual Growth

Mind maps can assist you in thinking more critically about your own life and objectives outside of the classroom and workplace. They work in two areas that are excellent and necessary for personal development: *sincere self-reflection and meticulous planning.*

Mind maps assist people in establishing *priorities, goals, and action plans*. They are motivational tools that help people stay focused and accountable by visualizing goals and progress.

Let's say you wish to be healthier. You could make a mind map with "Better Health" at the center and branches. Rather than making ambiguous promises to "get in shape," you will put in place the following items:

- *Exercise: the kind you like, the time you have, and the equipment you need*

- *Nutrition: meal planning, healthy substitutes, and present eating patterns*

- *Sleep: current trends, variables influencing the quality of sleep, and methods for improvement*

- *Stress management includes coping mechanisms, relaxation techniques, and stress sources.*

You can better understand how diet and exercise interact to determine overall health with this visual aid. And you may discover that getting too little sleep lowers your stress levels, which in turn increases your likelihood of eating junk food, which lowers your energy levels for exercise. Then, of course, it also can tend toward heart-related issues of concern. By arranging ideas and thoughts visually, mind mapping improves clarity and makes it simpler to prioritize values and set goals.

How to Begin Using Mind Maps

It's not as hard as you might think to make mind maps that work.

Beginning with a blank sheet of paper, take these easy actions:

1. Write your primary point in the middle of the page and draw a circle around it.

2. Sketch your primary subtopics as branches extending from the central circle. Yes, if at all possible, use different colors.

3. For more information and examples pertaining to each major subtopic, include smaller branches.

4. Look at the connections between different branches and draw lines to show how they relate to one another.

5. As you work, pose questions: What is absent? What connections exist between these concepts? What proof is there of this?

6. Keep it visual by emphasizing key points with symbols, images, or various writing styles. Using pencils or pens of different colors can sometimes help you highlight key portions of the image. You are the designer here.

There is no "right" way to create a mind map, so keep that in mind. Helping your brain recognize patterns and form connections is the aim. While some people like messy, imaginative maps, others prefer neat, well-organized ones. See what suits you.

A Novel Set of Skills

Not only is mind mapping a popular study method, but it's also an effective way to cultivate lifelong critical thinking abilities. Mind maps help you see the big picture while keeping track of crucial details, whether you're studying complex subjects, planning personal goals, making business decisions, or attempting to comprehend social issues.

It's important to keep in mind that mind mapping is about more than just information organization. It involves teaching your brain to question presumptions, seek connections, and tackle issues from several perspectives. Critical thinking is precisely characterized by these abilities.

Begin with a straightforward issue or choice you are currently dealing with. *Make a mind map and observe the connections that show up.* This visual way of thinking will become second nature to you with practice, enabling you to confidently and clearly navigate an increasingly complex world.

When you're starting something new, everyone needs a little help. **Having a template to start with can make a mind map creation**

much simpler. The following are some free ones that I found online that you can download and print.

Free Printable Mind Map Templates in PDF Format

TemplateLab.com

Templatelab.com is the domain.

Particular Templates on Hand:

Word, PowerPoint, PDF, and Photoshop templates for business plan mind maps

Template for an Essay Mind Map

Mind Map Template for Book Analysis

Mind Map Template for Character Development

Template for a Creative Mind Map

Template for a Bubble Mind Map

Template for a Brainstorming Mind Map

Template for a Personal Development Mind Map

Mind Map Template for Psychology

Template for an Advertising Mind Map

Template for a Career Planning Mind Map

Mind Map Template for Storytelling

Ideal For: Writers, students, business planners, and anybody else in need of organized mind maps for academic or professional purposes

Venngage

Web address: venngage.com

Particular Templates on Hand:

Templates for Strategy Planning

Templates for Campaign Planning

Templates for Workflow Visualization

Templates for Instructional Use

Downloads in Image and PDF Formats

Ideal for: Business strategists, educators, and marketing experts in need of aesthetically pleasing, editable templates

Template.net

Template.net is the domain.

Particular Templates on Hand:

Template for a Project Quality Mind Mapping Session

Mind Map Template for Short Stories

Templates for Personal Development

Templates for Business Planning

Mind maps in PDF format

Ideal For: Creative writers, project managers, and those interested in personal growth

Mindomo.com

Mindomo.com is the domain.

Particular Templates on Hand:

More than 20 distinct file formats (such as PowerPoint, PDF, and image)

Templates for Brainstorming

Templates for Project Planning

Study Guides

Templates for Businesses

Best Used For: Flexible format options for use in business, education, and personal projects

Biggerplate.com

Biggerplate.com is the domain.

Particular Templates on Hand:

Thousands of templates for businesses

Templates for education

templates unique to a given industry

Expert examples

Ideal for: Serious mind mappers searching for templates of the highest caliber in all sectors

Wondershare's EdrawMind Software

edrawmind.wondershare.com is the domain.

Particular Templates on Hand:

Templates for Project Management Mind Maps

Templates for Business Model Design

Templates for Book Analysis

Holiday/Halloween Templates

EdrawMind software's educational templates (EMMX files)

Ideal for: EdrawMind users looking for expert business and instructional templates

Canva.com

Canva.com is the domain.

Particular Templates on Hand:

Expert Mind Map Designs

Campaign Templates for Marketing

Templates for Education

Templates for Personal Development

Templates for Collaborative Teams

Ideal For: Groups in need of cooperative, eye-catching presentation templates for business use

Miro.com

Miro.com is the domain.

Particular Templates on Hand:

Templates for Business Mind Maps

Templates for Concept Maps

Map Templates for Empathy

Templates for Outcome Mappings

Templates for Mental Models

Templates for Business Plan Mind Maps

Ideal for: Collaborative brainstorming sessions, business strategists, and remote teams

Figma's FigJam

Figma.com is the domain.

Particular Templates on Hand:

Templates for Brainstorming

Templates for Setting Goals

Chart Templates for Organizations

Map Templates for Empathy

Templates for Project Planning

Templates for Debate Planning

Ideal for: Creative collaborators, UX specialists, and design teams

The XMind website

xmind.com is the domain.

Particular Templates on Hand:

20+ Templates Ready for Presentations

Templates for Business Meetings

Templates for Strategy Discussions

Templates for Timelines

Templates for Fishbone Analysis

Templates for Solving Problems

Templates for Organization Charts

Ideal for: Experts who require mind maps that are ready for presentations for meetings and strategic planning

MindMeister.com

Mindmeister.com is the domain.

Particular Templates on Hand:

Templates for Business Strategies

Templates for course syllabuses

Templates for Project Retrospectives

Templates for Project Planning

Templates for SCAMPER Brainstorming

The Templates for "How Might We"

Ideal for: Teams concentrating on structured brainstorming techniques, educators, and project managers

Clickup

Clickup.com is the domain.

Particular Templates on Hand:

Templates for Project Quality Assurance

Multipurpose Blank Templates

Templates for Project Planning

Integration Templates for Task Management

Ideal for: Teams and project managers who wish to incorporate mind maps into task management software

These resources provide templates that range from basic blank frameworks to intricate designs tailored to a particular industry. The majority of websites provide a variety of file formats (such as Word, PowerPoint, PDF, and others), so you can access them with any program you choose.

However, there are those of us who would rather work on our computers. A list of programs that are either free or free to try is provided here.

Free "mind mapping software" download for desktop computers running Linux, Mac, and Windows

You can make mind maps on your computer with a variety of great software applications, many of which are totally free. Here is a rundown of the top choices:

Free Software for Desktop Mind Mapping (nb: there may be changes)

The FreeMind

Cost: Open source and totally free

Platforms: Linux, Mac, and Windows

Qualities:

It is compatible with all major operating systems because it is written in Java.

Quick, easy-to-use interface with "fold/unfold" functions that only require one click

Export to HTML, PDF, and PNG formats

supports file attachments, notes, and hyperlinks.

Shortcuts on the keyboard for easy navigation

Compatibility across platforms

Ideal for: People looking for a simple, dependable mind mapping tool that has been validated over time

The Free Version of SimpleMind

Cost: The pro version costs $29.99 to $34.99, while the free version is available.

Platforms: iOS, Android, Mac, and Windows

Qualities:

Streamlined, contemporary interface

Options for free-form and auto-layout

To cut out distractions, use autofocus mode.

Syncing across platforms

Basic export features in the free edition

Ideal for: Consumers seeking a cutting-edge, intuitive interface that works on mobile devices

Free Plan Mindomo

Availability of a free plan (premium plans start at $4.50/month)

Platforms: Linux, Mac, Windows, and the web version

Qualities:

Both online and offline

Three mind maps are included in the free plan.

Excellent PDF export quality

Support for multimedia (pictures, videos, and audio)

Features for real-time collaboration

Various export formats

Ideal for: Users desiring occasional collaboration and professional features

The free version of EdrawMind

Cost: Pro version $59/year; free version available

Platforms: Linux, Mac, and Windows

Qualities:

Expertly designed templates

Timelines and Gantt charts

Mode of presentation

Integration of cloud storage

Brainstorming mode that generates automatically

Ideal for: Students and business users who require mind maps with a polished appearance

The GitMind

Cost: Nothing at all

Platforms: Web, iOS, Android, Mac, and Windows

Qualities:

Mind maps indefinitely

Collaboration in real time

Several templates

Export in a number of formats

Device synchronization across devices

Ideal for: Individuals and groups in need of limitless free mind mapping

Free Online Resources for Mind Mapping

The MindMup

Cost: Free for up to 100KB of maps

Platforms: Web-based (compatible with any internet-connected computer)

Qualities:

Registration is not necessary.

Integration with Google Drive

An infinite number of public maps

Export in a number of formats

Editing together

Ideal for: People who want instant access and don't want to download software

The Coggle

Up to three private maps are free.

Platforms: online

Qualities:

An infinite number of public maps

Support for markdown formatting

Simple collaboration and sharing

Clear and user-friendly interface

Exporting images and PDFs

Ideal for: Infrequent users who don't mind the map restriction

Miro's Free Plan

Cost: Up to three boards are included in the free plan.

Platforms: desktop apps and web-based

Qualities:

An endless canvas

Cooperation within the team

Frameworks and templates

Combining other tools

Mode of presentation

Ideal for: Groups engaging in cooperative planning and brainstorming

Options for Open Source

The Freeplane

Cost: Open Source and totally free

Platforms: Linux, Mac, and Windows

Qualities:

An upgraded edition of FreeMind

A more contemporary user interface

Options for advanced formatting

Support for formulas

Support for scripts in automation

Ideal for: High-level users seeking extensive personalization

View Your Mind, or VYM

Cost: Open source and totally free

Platforms: Linux, Mac, and Windows

Qualities:

Integration of note-taking

Priority and flag systems

Export in several formats

Management of links

Support for images

Ideal for: Individuals who wish to integrate thorough note-taking with mind mapping

Limited Free Versions of High-End Programs

The Free Version of XMind

The pro version costs $4.92 per month, while the free version is available.

Platforms: iOS, Android, Linux, Mac, and Windows

Qualities:

Expertly designed templates

Various forms of structures (fishbone, matrix, mind maps)

Basic export features in the free edition

A sleek, contemporary interface

Ideal for: Customers seeking high-quality maps with the ability to upgrade

Free Trial of MindMeister

Up to three mind maps are free.

Platforms: online

Qualities:

Tools for professional collaboration

Editing in real time

Mode of presentation

Available mobile applications

Ideal for: Expert teams wishing to test things out before committing to paid plans.

Suggestions for Each Use Case

For students, both **FreeMind** and **SimpleMind** Free provide good features at no cost.

For business users, consider **EdrawMind** or **Mindomo**, which offer professional features and upgrade options.

For teams, **MindMup** or **Miro** offer great collaboration tools.

For power users, **VYM** or **Freeplane**, which offers more features and customization

For easy use, *MindMup* or **Coggle** are quick, simple, and don't require a download.

Most of these applications are available for direct download from their official websites. You can begin using the totally free options, such as **FreeMind and SimpleMind**, right away and with no restrictions. Usually, the online tools only require the creation of a free account. You might be interested to know that *FreeMind was originally developed by a teacher* to help his students learn concepts. He uploaded it for free to help other teachers, students, and self-learners.

The great thing about having so many free options is that you can test out a few to determine which interface and feature set best suits your needs and way of thinking. *Before possibly moving on to more feature-rich options*, many users find that beginning with a basic tool like FreeMind or Coggle helps them understand mind mapping.

One thing to remember is that the internet is changing not daily, but within hours and minutes, so always know that what you're looking for may be at a different URL. Unfortunately, the costs of some of these programs have changed, but there are still sufficient free ones or those that offer free trials for everyone.

If you're a real computer nerd, I'm sure you know all about *GitHub* and what it offers. It's much more complex, to my way of thinking, but it offers an inordinate number of free programs that are added daily. In fact, I receive daily newsletters that contain at least three to four new programs that stretch the mind unbelievably. They're not mind maps, but they increase our ability to use AI in ways we may never have conceived. I always subscribe (they're free) and read them over carefully.

Chapter 15: When Stress Hijacks Your Mind

How to Protect Your Critical Thinking When Life Gets Overwhelming

Sarah felt her chest tighten as she read the urgent email from her boss at 9:47 PM. The project deadline had moved up by two weeks, her teenage daughter was struggling in school, and her elderly father had just been diagnosed with diabetes. Standing in her kitchen with dirty dishes piled in the sink, Sarah realized she couldn't think straight anymore.

The woman who normally analyzed problems methodically at work now found herself making snap decisions about everything. She'd agreed to volunteer for the school fundraiser without checking her calendar. She'd bought an expensive "miracle" supplement online after reading one testimonial. Most concerning, she'd snapped at her daughter for asking a simple question about homework.

Sarah didn't recognize herself. The stress wasn't just making her tired—it was hijacking her ability to think clearly. She was making choices that the "normal" Sarah would never make, and she felt like her mind was working against her instead of for her.

What Sarah was experiencing happens to millions of people every day. Stress doesn't just make us feel bad—it fundamentally changes how our brains process information, evaluate options, and make decisions. The good news? Understanding this process gives us the power to fight back.

How Stress Rewires Your Thinking Brain

When you're stressed, your brain doesn't just feel different—it actually functions differently. The prefrontal cortex, where your critical thinking happens, gets less blood flow and energy. Meanwhile, your amygdala, the brain's alarm system, becomes hyperactive and starts calling the shots. This portion of your brain is ready to fire away at a moment's notice and not stop.

Think of it this way: your brain has a CEO (prefrontal cortex) who normally makes thoughtful, strategic decisions. But when stress hits, it's like the security guard (amygdala) bursts into the boardroom, throws papers everywhere, and starts making all the decisions based on fear and urgency.

This biological hijacking shows up in predictable ways. You might find yourself buying things you don't need because they promise quick solutions, or saying yes to commitments without thinking them through. I knew a woman whose husband was a successful professional, and who had been diagnosed with ALS. In effect, this was a death sentence, and she began sitting in her living room and constantly ordering from online programs. She never opened the packages, and they began piling up. That was the reason that they went for therapy. It was more than alarming news, and she couldn't handle it, so she tried to soothe her stress by buying things.

When this happens, you start believing alarming news without checking sources, and you make important decisions based on how you feel in the moment rather than what makes logical sense. The solutions that would normally be obvious become nearly impossible, and you begin taking everything personally, even neutral comments from colleagues or family members. You fly off the handle frequently.

The most insidious part is that this creates a self-perpetuating cycle. Stress reduces critical thinking, which leads to poor decisions that create more problems. Those additional problems generate more stress, which further compromises your thinking ability, leading to even worse decisions. Before you know it, you're trapped in a downward spiral where each day seems to bring new crises that you feel less and less equipped to handle.

The Invisible Ways Stress Sabotages Your Judgment

Information processing becomes chaotic when stress floods your system. Your brain starts looking for quick answers instead of good ones, much like someone trying to read during an earthquake. Instead of carefully considering alternatives, you accept the first solution you hear. Rather than seeing possibilities, you become fixated on problems. What used to be neutral situations now appear threatening, and you find yourself making decisions based on worst-case scenarios while struggling to remember important details that would normally come easily.

Maria's experience illustrates this perfectly. She normally researched major purchases carefully, reading reviews and comparing prices across multiple sources. But when her car broke down during a particularly stressful period at work, the careful analyst disappeared. She bought the first replacement she saw at a dealership, paying $3,000 more than necessary and ending up with a car that didn't meet her actual needs. The stress had effectively turned off her ability to think strategically about a significant financial decision.

The *confirmation bias trap* becomes even deeper under stress. When your nervous system is activated, you become more likely to seek information that confirms what you already believe, especially if those beliefs make you feel safer or more in control. You might find yourself only reading news that matches your existing opinions, dismissing expert advice that contradicts what you want to hear, or surrounding yourself with people who agree with you. Every neutral event gets interpreted through the lens of either threat or salvation, with little room for nuanced understanding.

Decision fatigue transforms into decision paralysis when you're already running on empty. In fact, I recently wrote a book (*When You Can't Pour From an Empty Glass: CBT Skills for Exhausted Caregivers)* that contained information needed for caregivers who are truly "running on empty." Every choice you make throughout the day uses mental energy, and when you're stressed, even small decisions can feel overwhelming. You might spend thirty minutes choosing what to eat for lunch, avoid important decisions entirely, or make choices based purely on what requires the least effort. Sometimes you find yourself letting other people make decisions for you, not because you trust their judgment more than your own, but because you simply don't have the energy left to think through the options.

The Hidden Stress Triggers That Wreck Your Thinking

Information overload has become one of the most pervasive yet under-recognized sources of chronic stress in modern life. The constant stream of notifications, news updates, social media posts, and emails creates a state where *your brain never gets a chance to rest and reset*. Every ping represents a tiny dose of stress that accumulates throughout the day, leaving you feeling frazzled even when nothing particularly dramatic has happened.

This manifests in ways that many people don't connect to information overload. You might check your phone first thing in the morning and immediately feel anxious about the day ahead. Despite being caught up on your work, you feel perpetually behind on everything. Your attention span shrinks to just a few minutes at a time, and you find yourself making impulsive responses to emails or texts that you later regret. (Remember that 5-minute rule I mentioned earlier?)

The perfectionism trap creates enormous stress while ironically making good decision-making impossible. When you demand perfect decisions from yourself, you create a mental environment where clear thinking can't flourish. This shows up as spending hours researching minor decisions, feeling paralyzed by important choices, redoing work that's already good enough, and focusing more energy on avoiding mistakes than on achieving meaningful goals.

Comparison stress operates more subtly but just as destructively. *Constantly measuring yourself against others*—whether through social media, workplace competition, or family dynamics—creates chronic stress that clouds judgment in predictable ways. You start making decisions based on what looks good to others instead of what actually works for you. You underestimate your own capabilities while overestimating what others can do. Are you a victim of the *"imposter syndrome?"* What is that?

The *feeling of being a fake* despite your actual achievements defines imposter syndrome. You experience persistent anxiety about others discovering your lack of understanding while believing *your achievements stem from chance* instead of your genuine abilities and dedication.

People who experience imposter syndrome typically believe they don't deserve their place or that others misjudge their intelligence. These individuals minimize their accomplishments while constantly fearing that others will discover their supposed lack of competence. The phenomenon affects individuals at every level, from students to CEOs, because it *occurs frequently.*

The good news? The process of recognizing these feelings marks the beginning of overcoming them. Most successful individuals have experienced these feelings at some point in their lives, and learning to trust your abilities and acknowledge your real accomplishments helps you **silence your inner critic**.

Simple Home-Based Solutions That Actually Work

The key to protecting your critical thinking isn't about eliminating stress entirely—that's neither possible nor necessarily desirable. Stress can motivate action and signal that something needs attention. The goal is to *create small pockets of calm* that allow your thinking brain to come back online when you need it most.

The *90-second reset* serves as a practical circuit breaker for the stress-thinking cycle. When you feel stress starting to hijack your thinking, begin by noticing where you feel it in your body. Stress manifests physically before it completely takes over your mind—tight shoulders, clenched jaw, shallow breathing, or tension in your stomach.

Once you've identified the physical sensation, *take three slow breaths, making your exhale longer than your inhale.* Count "in for 4, out for 6" to give your mind something concrete to focus on. Then ask yourself one simple question: "*What's one thing I can control right now?*" Focus only on that single element within your influence.

This technique works because *it takes approximately 90 seconds for stress hormones to cycle* through your system when you're not continuously adding fuel to the fire. It isn't meditation or complex mindfulness practice—it's a practical reset button that you can use before responding to that urgent email, before making that purchase decision, or before having that difficult conversation. Those 90 seconds can mean the difference between a reaction you'll regret and a response you'll be proud of.

Creating a thinking corner in your home provides a physical anchor for clearer decision-making. This designated space doesn't need to be elaborate—a comfortable chair by a window, a spot at your kitchen table with a notepad, or even a corner of your bedroom with a pillow on the floor. The key is *consistency and intention*. When faced with a decision that matters, go to your thinking corner and *spend 5–10 minutes just sitting with the question*. Don't force solutions or pressure yourself to reach conclusions. Instead, let your mind wander around the problem naturally. Look out the window at the clouds and try to make faces or shapes out of them as a distraction. It works.

Your brain needs physical cues to shift into different modes of operation. Having a designated thinking space trains your mind to slow down and consider rather than react. Keep a small digital tape recorder or a notebook in this space for jotting down thoughts—not necessarily decisions, just the ideas that emerge when you give yourself permission to think without immediate pressure to act.

The *daily stress audit* creates awareness without judgment about how stress affects your thinking patterns. Each evening, spend five minutes reviewing your day through the lens of stress and decision-making. Consider when you thought clearly and what decisions you're glad you made. Reflect on when you felt calm and in control versus when stress affected your judgment. Think how you might act differently in the same situations and what caused your stress.

Most important, use this information to plan for tomorrow. Ask yourself how you can create more moments of clear thinking and what stress triggers you can anticipate and prepare for. This isn't about judging yourself harshly for imperfect decisions—it's about preparation, gathering data on your own patterns so you can work with them instead of against them.

Building Your Stress-Resistant Thinking System

Creating decision rules in advance protects you from having to make important choices when your judgment is compromised. When you're calm and thinking clearly, establish simple guidelines for common stressful situations. These might include waiting 24 hours before making any purchase over $100 when you're feeling stressed, saving difficult emails as drafts instead of responding immediately, or telling people who pressure you for quick answers that you'll think about it and get back to them tomorrow. There's no harm in putting things off a bit rather than acting quickly, impulsively, and not in your best interests.

The key is recognizing that your future stressed self will not have access to the same clear thinking you have right now. By creating these guidelines when your judgment is sound, you're essentially leaving helpful instructions for yourself to follow when stress inevitably clouds your thinking again. If you ever want to gauge your stress level, there are places on the internet that have the reliable *Holmes and Rahe Stress Index*. It will assign a number to all your activities, showing when and where you're most stressed and vulnerable.

Developing a trusted advisor system provides external perspective when your internal compass gets disrupted by stress. Identify two or three people who can serve as thinking partners—not people who will make decisions for you, but individuals who ask good questions and help you see situations more clearly. These should be people who don't have strong opinions about your specific choices but are skilled at helping you think through problems systematically.

The most effective trusted advisors help you access your own wisdom rather than imposing their judgment on your situation. When you call them during stressful periods, ask them to help you think through problems rather than solve them for you. Request that they point out when you seem to be thinking from stress rather than clarity, and be open to their observations about patterns they notice in your decision-making.

Energy management proves more effective than time management when it comes to protecting your critical thinking abilities. Your capacity for complex reasoning fluctuates throughout the day *based on your physical and mental energy levels*. Instead of trying to force important decisions during low-energy periods, learn to recognize when your thinking resources are abundant versus depleted.

Schedule significant decisions for times when your energy is naturally higher, and recognize when you're running on empty so you can avoid major choices during those periods. Use high-energy times for complex thinking tasks while saving routine decisions for when your mental resources are lower. Pay attention to signs that your thinking energy is compromised—everything feels harder than it should, you're avoiding decisions instead of making them, small problems feel overwhelming, or you're more irritable and emotional than usual.

The Physiology of Clear Thinking

Your brain operates as part of your body, and your body's condition directly affects your thinking abilities. You don't need expensive equipment or gym memberships to support your brain's optimal performance—simple, accessible approaches can make a significant difference.

Walking, especially outdoors, serves as one of the most powerful tools for clearing stress and restoring mental clarity. Research demonstrates that even a 10-minute walk can improve cognitive function *for up to two hours afterward.* The key is using walking strategically rather than just for exercise. Sometimes walk without podcasts or music, allowing your mind to wander freely. Use walking time to process decisions rather than escape from them. Pay attention to your surroundings, as this naturally calms the stress response and helps your nervous system reset.

When you're feeling stuck on a problem or overwhelmed by a decision, walking often provides the mental space needed for solutions to emerge. The rhythmic nature of walking, combined with the gentle stimulation of changing scenery, creates ideal conditions for the kind of loose, associative thinking that leads to insights and creative problem-solving.

Your breathing pattern, too, directly influences your nervous system and thinking clarity. When stressed, most people shift to shallow, rapid breathing, which maintains the stress response and keeps clear thinking offline. The 4-7-8 breathing technique—inhaling through your nose for 4 counts, holding for 7 counts, and exhaling through your mouth for 8 counts—activates your parasympathetic nervous system, which is necessary for calm, rational thinking.

This isn't about becoming a breathing expert or spending long periods in meditation. It's about having a reliable tool you can use before important conversations, decisions, or any time you notice stress beginning to affect your judgment. Three or four cycles of breathing can shift your nervous system *from stress mode to thinking mode in less than two minutes.*

The connection between sleep and thinking goes far beyond simple fatigue. Poor sleep specifically impairs the prefrontal cortex functions you need for critical thinking, decision-making, and emotional regulation. Protecting your sleep quality directly protects your ability to think clearly during challenging situations.

Simple sleep improvements that support better thinking include *keeping your bedroom cool (around 65-68°F), stopping screen use at least one hour before bed, and using your bedroom only for sleep* rather than work or entertainment. If you frequently wake up in the middle of the night with racing thoughts or worries, keep a notebook by your bed to quickly write down concerns so your mind can let go of them until morning.

Stress-Proofing Your Information Diet

Constant news consumption creates chronic stress that impairs judgment, yet many people consume information continuously without recognizing this connection. Setting specific boundaries around news and information can dramatically reduce background stress levels and improve decision-making capacity.

Rather than checking news continuously throughout the day, establish scheduled times for information consumption—perhaps 15-20 minutes in the morning and 15 minutes in the evening. Choose two or three reliable news sources instead of trying to consume everything available. Avoid news entirely during your first hour awake or your last hour before sleep, as these are critical times for setting your nervous system's tone for the day or night.

Social media platforms are specifically designed to capture attention and create emotional responses—exactly what interferes with clear thinking. Rather than trying to eliminate social media entirely, create boundaries that protect your mental resources. *Remove social media apps from your phone's home screen* so accessing them requires intentional effort rather than reflexive checking. Use built-in app timers to limit daily usage, and pay attention to how different accounts and types of content affect your mood and stress levels.

Consider unfollowing accounts that consistently make you feel stressed, inadequate, or agitated, regardless of whether their content is "important" or "educational." Your mental resources are finite, and protecting them from unnecessarily stressful content is a *form of self-care* that directly supports better thinking.

When information overload becomes particularly intense, try implementing a temporary information fast. Choose one day per week to avoid all non-essential information—no email checking, news reading, or social media browsing. Use this time for reflection, rest, or activities that bring you genuine enjoyment. Notice how your thinking changes when you're not constantly consuming and processing new information. In my area of healthcare, decades ago we incorporated the idea of a "mental health day" into our schedules. Everyone benefited from it, and no one had to feel guilty because they were calling in "sick."

Building Long-Term Resilience

Developing a growth mindset toward stress transforms it from something that happens to you into information about what matters to you and what might need to change in your life. Instead of viewing stress as purely negative, begin seeing it as data about your values, priorities, and capacity. Know that stress comes from everything—whether it's good news or bad news. Planning a wedding, going on vacation, or even getting a promotion are all stressful, but we deal with them.

When you catch yourself thinking "*I can't handle this,*" try reframing to "*This is challenging, but I can figure it out.*" Replace "*Everything is falling apart*" with "*Some things are difficult right now.*" Transform "*I always make bad decisions under pressure*" into "*I'm learning to make better decisions when stressed.*" Have you ever heard the song from "*You're a Good Man, Charlie Brown*," where he talks about stress? He says, "*I always think best under pressure. So if I wait till tomorrow cause there'll be lots of pressure if I wait 'til tomorow, so I'll wait til tomorrow.*"

It's partially good and partially bad. Stress can be, as I have said, both good and bad. Procrastination is not great, and sometimes we do need to act in the moment and not put it off till tomorrow. But sometimes we also have to wait until tomorrow. These aren't empty positive affirmations—they're more accurate descriptions of reality that leave room for growth and learning. It's putting reframing to good use for yourself.

Stress inoculation involves gradually exposing yourself to manageable amounts of stress to *build your tolerance and thinking skills under pressure*. This doesn't mean seeking out unnecessary stress, but rather using the naturally occurring challenges in your life as opportunities to practice your stress-management and clear-thinking techniques.

When facing mildly stressful situations, deliberately practice your thinking strategies rather than just trying to get through the experience. Notice what helps you maintain clarity when the pressure is moderate, and gradually build your tolerance for uncertainty and discomfort. Like physical exercise, this kind of *mental training* strengthens your capacity over time.

Having clear personal values serves as an anchor during stressful periods when emotions and external pressures might otherwise drive your decisions. Take time to identify what matters most to you when everything else is stripped away—what principles you want to guide your decisions, how you want to be remembered, and what kind of person you want to be when things get difficult.

When stress clouds your judgment, these values can *serve as a decision filter*. Ask yourself what someone with your values would do in this situation, and choose the option that aligns with your long-term character goals rather than your immediate emotional state. Values-based decisions usually feel right even when they're difficult, and they're decisions you're less likely to regret later.

Your Personal Stress-and-Thinking Action Plan

Building stress resilience and protecting your critical thinking abilities work best as a gradual process rather than a dramatic overhaul. Start by spending *one week simply building awareness of how stress affects your thinking patterns*. Keep a simple log of when stress influences your decisions, and notice your physical stress signals like tight shoulders or shallow breathing, Even stomach upset. Identify your most common stress triggers, and practice the 90-second reset technique at least once daily.

During the second week, *focus on environmental design*. Create your thinking corner and begin using it for important decisions. Establish boundaries around news and social media consumption that feel sustainable rather than restrictive. Set up your trusted advisor system by identifying people who can serve as thinking partners. Begin the daily stress audit practice, spending just five minutes each evening reviewing how stress affected your day.

The third week *emphasizes skill building*. Implement the two-list method for current challenges you're facing, separating what you can control from what you can't. Practice walking as a thinking tool, using it strategically when you feel stuck or overwhelmed. Try the breathing technique when you notice stress beginning to affect your judgment. Create your first set of decision rules for common stressful situations.

By the fourth week, *focus on integration and long-term planning*. Hold your first kitchen table strategy session, giving yourself uninterrupted time to think about your priorities and challenges. Assess which techniques work best for your personality and lifestyle. Plan how to maintain these practices long-term without making them feel like additional pressure. Identify areas where you'd like to continue growing your stress resilience.

Chapter 15 Skills Checkpoint

Before moving forward, ensure you can recognize stress signals and identify when stress is affecting your thinking before making important decisions. Practice using the 90-second reset to interrupt the stress-thinking cycle when it begins. Implement environmental supports by creating physical and digital boundaries that protect your thinking space.

Develop energy management skills so you can make important decisions when your thinking resources are most abundant. Apply stress-resistant decision-making techniques like the two-list method and predetermined decision rules to maintain good judgment under pressure.

Consider these self-assessment questions: What are your most common stress triggers, and how do they specifically affect your thinking patterns? Which of the home-based techniques from this chapter feels most doable for your current lifestyle? How can you tell the difference between stress-based thinking and clear thinking in your experience?

For your real-world application challenge, focus on one stressful area of your life over the next week and apply three techniques from this chapter. Choose one environmental change, such as creating a thinking corner or establishing information boundaries. Practice one in-the-moment technique like the 90-second reset, strategic breathing, or walking for clarity. Implement one planning technique, such as decision rules, values-based anchoring, or trusted advisor consultation.

Document how these changes affect both your stress levels and your decision-making quality.

Like Sarah learned after implementing these techniques, you can experience stress without letting it control your mind. You can face challenging situations while maintaining access to your wisdom, creativity, and judgment. And you have more control over your thinking than you might realize, and that control becomes strongest precisely when you need it most.

Remember one thing: No matter what anybody tells you about what other people are doing or what research has said, as one of my mentors (a neuropsychologist) told me during my internship, *"Everyone is a sample of one, and we can't apply everything in research to them."* You are more unique in ways you have yet to discover. So accept that, use it when you can, carefully think those aspects of yourself over, and then proceed.

Your enhanced stress-management skills now prepare you to apply all your critical thinking abilities consistently, regardless of external pressures or internal turbulence. **You're building not just better thinking skills**, but a **more resilient and capable version of yourself.**

Chapter 16: The Lost Art of Letting Your Mind Wander

Tom felt guilty sitting on his back porch at 2 PM on a Tuesday, watching clouds drift across the sky while his laptop waited inside with seventeen unread emails. His neighbor was mowing the lawn, his wife was in back-to-back Zoom calls, and here he was, doing absolutely nothing productive. He should work, plan, optimize, and achieve something—anything apart from just sitting there like a lazy person. Yes, he was feeling very guilty indeed.

But something intriguing started happening as he watched those clouds. His mind began connecting ideas that had been floating around separately for weeks. The marketing problem at work suddenly linked to something his daughter had said about her friends' shopping habits. A conversation with his brother about their aging parents sparked an insight about the client presentation he'd been struggling with. By the time he finally went back inside, Tom had solved three problems that had been nagging at him for days, and he felt more mentally refreshed than he had in months.

He had stumbled onto something that our productivity-obsessed culture has largely forgotten: **the profound value of letting your mind wander.** In our rush to fill every moment with purpose and efficiency, we've lost touch with the kind of loose, undirected thinking that actually fuels creativity, problem-solving, and mental well-being.

The Tyranny of Constant Productivity

We live in a world that treats an idle mind like a character flaw. Every moment must be optimized, every minute accounted for, and every thought directed toward a specific goal. We listen to podcasts while walking, check emails while waiting in line, and feel guilty if we're not learning something, improving something, or accomplishing something.

This relentless focus on productivity creates a peculiar mental exhaustion. Your brain becomes like a computer with too many programs running simultaneously—everything slows down, creativity diminishes, and even simple problems feel overwhelming. We've convinced ourselves that constant mental activity equals progress, but research suggests the opposite might be true.

The default mode network in your brain—the neural circuits that activate when you're not focused on specific tasks—plays a crucial role in memory consolidation, self-reflection, and creative insight. Constantly being "on," directed, and productive, essentially deprives your mind of its most valuable functions.

Think about the last time you had a genuine breakthrough or creative idea. Chances are, it didn't happen during a focused work session or while you were trying hard to solve a problem. It probably occurred in the shower, on a walk, while doing dishes, or in some other moment when your conscious mind was occupied with something simple and your deeper mental processes had space to work.

I know an orthopedic surgeon who has published several books on religion and medicine, who revealed that the shower is one of his most creative places. To capture all of his creative effort that comes, eventually to fruition while showering, he has a special waterproof tape recorder that he places there. Each day when he takes a shower, the tape recorder is on and captures everything that he comes up with. It has been a truly successful effort for him.

What Happens When You Let Your Mind Roam

Mind wandering isn't laziness—it's a sophisticated cognitive process that allows your brain to make connections it simply cannot make when you're focused on specific tasks. When your attention isn't locked onto immediate concerns, your mind sorts through information, linking seemingly unrelated concepts and processing experiences in ways that lead to insights and solutions.

During these periods of mental freedom, your brain doesn't shut down—it shifts into a different mode of operation. Memories are reorganized, patterns emerge from chaos, and creative possibilities that were invisible during focused work suddenly become apparent. The marketing executive who gets her best ideas during her morning jog isn't being unproductive—she's allowing her brain to work in the way it evolved to work.

Sarah, a software designer, discovered the phenomenon accidentally when she started taking her lunch breaks in a small park near her office instead of eating at her desk while working. Initially, she felt guilty about the "wasted" time, but she soon noticed that her afternoon coding sessions became more creative and efficient. Problems that had seemed intractable in the morning often resolved themselves after she'd spent thirty minutes watching people walk their dogs and letting her thoughts drift wherever they wanted to go.

The Creative Power of Purposeless Time

Creativity requires a delicate balance between focused effort and unfocused wandering. You need concentrated work to develop skills, gather information, and execute ideas, but you also need unstructured mental time for those ideas to germinate and connect in unexpected ways.

Many artists and writers have long understood this principle. They'll work intensively on a project, then deliberately step away and do something completely unrelated—gardening, cooking, walking, or simply staring out the window. This isn't procrastination; it's an essential part of the creative process. I once interviewed Isaac Asimov, a prolific author who (at that time) had written 125 books. I asked him how he did it, and his answer was simple: *"Whenever I become tired of a project or I feel I can't go on, I put it back in the drawer and take a different one so that my mind is fresh and I feel I can proceed in a renewed way."* He undoubtedly comprehended the significance of shifting activities and releasing himself from monotony as a means of liberation. In fact, he had nine projects, each in a separate drawer, in his desk. He showed me some of them, one by one, and they went from chemistry to Shakespeare and everything in between.

The key insight is that your subconscious mind continues working on problems even when your conscious attention is elsewhere. In fact, it often works better when you're not trying to force solutions. This phenomenon is known as **the shower effect**. The pressure and narrow focus of deliberate problem-solving can actually prevent you from seeing possibilities that become obvious once you relax your mental grip.

Reclaiming the Right to Mental Downtime

Learning to value mental wandering requires overcoming decades of conditioning that equates busyness with worthiness. You might need to give yourself explicit permission to do "nothing," recognizing that *this nothing is actually something quite valuable.*

Start small and practical. Instead of scrolling through your phone while waiting for appointments, try just sitting and letting your mind drift. When you're walking from one place to another, resist the urge to make phone calls or listen to content—use that time for unstructured thinking. **Create pockets of purposeless time in your day**, even if they're just five or ten minutes long.

The shower room, oddly enough, often serves as one of the last refuges for mind wandering in our hyper-connected world. Many people say they have their best ideas (like the surgeon) there, not because of the location, but because it's one of the few places where we're not multitasking or consuming information.

Working With Your Natural Rhythms

Your capacity for focused work and creative thinking fluctuates throughout the day in predictable patterns. Instead of fighting these natural rhythms, you can learn to work with them, using high-focus periods for concentrated tasks and low-focus periods for mental wandering. I saw that taking time between 12:00 and 3:00, as they used to do in Europe, was a good thing. It broke the day up from the usual work activities and permitted that natural cycle during the day to have its time to refresh and relax. During my first trip to Europe, I was completely unaccustomed to the practice of taking breaks, as everything in the United States is open almost all of the time. When I saw that the restaurants were closing at 2 and the banks closed at noon, it was startling. Then I understood.

Most people experience a creativity peak in the late morning, a focus dip in the early afternoon, and another creative surge in the early evening. Rather than forcing yourself to power through the afternoon slump with more coffee and determination, that might be the perfect time to take a walk, sit outside, or engage in some other activity that allows your mind to roam freely.

Pay attention to when your best ideas typically occur. If you're a morning person who gets insights during early walks, protect that time and use it intentionally. If you're a night owl whose mind opens up during evening downtime, honor that pattern instead of forcing yourself into someone else's productivity schedule.

The Art of Productive Daydreaming

There's a difference between mindless distraction and purposeful mind wandering. Scrolling through social media might feel like letting your mind wander, but it's actually keeping your attention captured by external stimuli. True mind wandering happens when your attention is free to move wherever it wants to go, without external inputs directing the flow.

Productive daydreaming often involves gentle, repetitive activities that occupy just enough of your conscious attention to quiet the goal-oriented part of your mind. Walking, especially in natural settings, provides the ideal conditions. The rhythm of your steps, the changing scenery, and the mild physical activity create a perfect environment for creative thinking.

All of this also gives me pause and a new appreciation for why adult coloring books have proven so popular. Have you ever thought about buying a coloring book? What do you think about it? It's actually a very good way to simply allow yourself to slow down and relax.

Other activities that promote beneficial mind wandering include gardening, cooking simple meals, doing crafts or puzzles, taking baths, or sitting by water. The common element is that these activities are engaging enough to prevent anxiety or restlessness but not so demanding that they prevent your mind from wandering.

Making Peace with Apparent Unproductivity

The hardest part of embracing mind wandering is often dealing with the guilt and anxiety that arise when you're not visibly accomplishing anything. In a culture that *measures worth through output*, as I've said, sitting quietly and thinking can feel selfish or lazy.

Reframe these moments as essential maintenance for your most important tool—your mind. Just as you wouldn't expect a car to run well without regular maintenance, you can't expect your brain to function optimally without periods of rest and restoration. The insights, solutions, and creative ideas that emerge from mind wandering often prove more valuable than anything you could have produced through forced effort. if you're feeling a bit guilty and unsure of it, do some searching on the internet for articles on mind wandering and the benefits it can have for you.

Consider keeping a simple record of insights or ideas that come to you during unstructured time. This isn't to turn mind wandering into another productivity hack but to help you recognize the value of these periods when your internal critic questions their worth. You also want to capture those quick snippets of ideas that come to you. Please jot it down in a small pocket notebook for future reference.

Your Mind Wandering Practice

Start by identifying natural opportunities for mind wandering that already exist in your day. Your commute, morning coffee, evening walk, or time spent in line at the store can all become opportunities for purposeless thinking. The goal isn't to add more activities to your schedule but to stop filling existing gaps with stimulation and instead allow them to remain open.

Create one daily ritual that invites mind wandering. Sit outside for ten minutes each morning, take a shower without rushing, or spend a few minutes looking out your window before starting work. The specific activity matters less than your intention to let your mind roam freely during that time.

When you notice your mind starting to wander during focused work, don't immediately pull it back. Sometimes this is your *brain's way of telling you it needs a break* or that you're missing something important. Take note of where your thoughts were heading—you might discover that your mind was actually working on exactly what you need to be thinking about.

Remember that mind wandering is **not a luxury or a sign of laziness**—it's a fundamental human capacity that contributes to mental health, creativity, and problem-solving. In a world that demands constant focus and productivity, learning to let your mind wander freely becomes an act of both self-care and rebellion.

Tom discovered that his afternoon cloud-watching sessions weren't time stolen from productivity—they were investments in a different kind of productivity, one that valued insight over output and creativity over efficiency. **Your wandering mind might just lead you to exactly where you need to go.**

Chapter 17:
Don't Just Follow Trends: Think Critically About Your Career Future

The bottom line: Healthcare, technology, and green energy jobs will **see the biggest growth through 2050.** But the smart job seeker doesn't just follow trends—they use critical thinking to evaluate the evidence and ask the right questions about what skills will make them valuable no matter how the economy changes.

Every day, you're bombarded with claims about which careers are "hot" and which ones are "dead." Social media influencers promise that learning to code will guarantee your future. News articles warn robots will take everyone's jobs. Career counselors push you toward whatever field is trending this year. But here's the problem: most people accept these claims without asking the most important questions a critical thinker should ask.

Where does this information come from? Who conducted the research, and what methods did they use? Are multiple reliable sources saying the same thing? What assumptions are built into these predictions? And most importantly—what evidence would prove these predictions wrong?

When you're choosing a career path that could last 30 or 40 years, you can't afford to make decisions based on hype, fear, or outdated advice. You need to think like a detective, gathering evidence from multiple credible sources and weighing that evidence carefully. The good news? When you apply critical thinking skills to career planning, you're doing what several major research organizations that study job trends for a living are doing. All of them point to the same promising areas for long-term growth. But they also reveal important nuances that the headlines miss.

The Big Picture: What's Driving Job Growth

Before we dive into specific jobs, let's understand what forces shape the job market. Think of these as the currents in a river—if you know which way they're flowing, you can navigate much better.

The aging population is the biggest driver of job growth in America. The Bureau of Labor Statistics and World Economic Forum both identify healthcare and care services as the fastest-growing sectors, primarily because people are living longer and need more medical care. Almost 40 percent of the population was projected to be age 55 or older by 2024, compared with 34.2 percent in 2014, so we'll *need millions more people to care for an aging society.*

Technology and artificial intelligence are reshaping every industry. But here's the key insight most people miss: Unlike previous automation that mainly affected blue-collar work, AI is now disrupting cognitive and non-routine tasks in middle- to higher-paid professions. This process creates both challenges and opportunities. While some jobs disappear, new ones emerge that require people to work alongside smart machines.

Climate change and the green transition are creating entirely new job categories. It's these green transition trends that will drive growth and create 34 million additional jobs by 2030, from solar panel installers to environmental engineers. These aren't just feel-good jobs—they pay well and have strong growth projections.

Economic uncertainty and global changes are making companies value adaptable workers more than ever. Skill gaps are companies' biggest barriers—much more restrictive than regulatory issues or investment capital. This means that the workers who can learn and adapt will have the advantage.

The Jobs with the Strongest Growth Potential

Based on analysis from the Bureau of Labor Statistics, McKinsey Global Institute, World Economic Forum, and other major forecasting organizations, here are the job categories expected to see the strongest growth through 2050:

Healthcare: The Clear Winner

Home health and personal care aides are projected to grow 20.7% and add 820,500 jobs by 2033. But healthcare growth goes far beyond just nursing. Jobs include medical technicians, physical therapist assistants, occupational therapy assistants, and healthcare support specialists. Many of these positions require certificates or two-year degrees rather than four-year college programs.

The beauty of healthcare jobs is that they can't be outsourced or easily automated. People need human touch and care. Plus, as medical technology advances, it creates more healthcare jobs rather than eliminating them.

Technology and Data Jobs

Big data specialists, financial tech engineers, and AI/machine learning specialists lead growth in the fastest-growing tech positions. But you don't need to be a programmer to benefit from this trend. Technological skills are projected to grow in importance more rapidly than any other skills in the next five years, including basic digital literacy, cybersecurity awareness, and data analysis.

The key is understanding that "tech jobs" now exist in every industry. A farmer using GPS and data analytics, a factory worker operating computerized equipment, or a retail manager using inventory software—all of these require technological skills.

Green and Renewable Energy

Wind turbine service technicians and solar photovoltaic installers are projected to be the fastest growing occupations over the next decade. While these specific jobs have relatively small numbers, they represent a much larger trend toward environmental careers.

Green jobs include traditional roles like construction workers building energy-efficient buildings, as well as new positions like sustainability coordinators and environmental compliance specialists. Farm workers top the table of the largest growing jobs, partly due to increased focus on sustainable agriculture and food security.

Professional and Business Services

As the economy becomes more complex, companies need more specialized help. This includes roles like business analysts, project coordinators, training specialists, and customer success managers. Many of these jobs value communication skills and problem-solving ability over specific technical training.

The Jobs to Approach with Caution

Just as important as knowing what's growing is understanding what's shrinking. Postal service clerks, bank tellers, and data entry clerks have the greatest decline. Clerks could decrease by 1.6 million jobs, in addition to losses of 830,000 for retail salespersons, 710,000 for administrative assistants, and 630,000 for cashiers.

The pattern is clear: jobs involving repetitive tasks, data collection, and basic data processing are vulnerable to automation. If your job mainly involves following the same process over and over, it's worth considering how to add more complex, human-centered skills to your role.

What This Means for Your Career Strategy

Understanding job growth projections is just the first step. The real question is, how do you position yourself to benefit from these trends?

Focus on skills that complement technology rather than compete with it. Creative thinking and resilience, flexibility and agility are rising in importance, along with curiosity and lifelong learning. These are uniquely human capabilities that become more valuable as technology handles routine tasks.

Look for opportunities to develop "hybrid" skills. The most valuable workers will be those who understand both their core field and how technology can enhance it. A nurse who understands health data analytics, a construction worker trained in green building techniques, or a customer service representative skilled in using AI tools will have significant advantages.

Plan for continuous learning. Nearly 40% of skills required on the job are set to change by 2030. This doesn't mean your career is doomed—it means successful workers will be those who embrace ongoing education and skill development. Education didn't stop when you graduated from high school or college because you are going to be learning, potentially taking courses, even getting more certificates for your career.

Consider geographic factors. Job growth isn't uniform across the country. Research where your chosen field is growing strongest and consider whether relocation might be part of your strategy.

Making Smart Decisions in an Uncertain World

The most important skill for future job seekers might be the ability to think critically about conflicting information and uncertain predictions. No forecasting organization claims to predict the future perfectly. The Employment Projections program estimates specific values for projected employment levels and growth rates. However, this precision in the data does not account for the inherent uncertainty of predicting long-term changes in the labor market.

What this means is that you should *use these projections as a starting point* for your thinking, not as a guarantee. The smartest approach is to build a career that can adapt to multiple scenarios. Focus on developing strong fundamental skills in communication, problem-solving, and learning itself. Choose fields where you can see multiple pathways for growth rather than betting everything on a single prediction. Mind maps might be useful here.

The future job market will reward those who can navigate uncertainty, learn continuously, and add distinctly human value in an increasingly automated world. By understanding the major trends and positioning yourself thoughtfully, you can build a career that thrives regardless of exactly how the future unfolds.

The jobs of 2050 may look different from how we imagine them today. We know that the workers who will succeed in getting them are already developing the skills and mindset they'll need. The question is, **will you be one of them?**

References

Altstaedter, Laura Levi, and Lily Johnson. (2025). Fostering Learner Engagement in Study Abroad Contexts through Digital Mapping. "Innovation in Language Learning and Teaching," May, 1–11.

Ariely, D. (2008). "Predictably Irrational: The Hidden Forces That Shape Our Decisions." HarperCollins.

Aura. (2025). Job Market Analytics: Predict Industry Trends Before They Happen. Retrieved from https://blog.getaura.ai/job-market-analytics

Baird, Benjamin, Jonathan Smallwood, Michael D. Mrazek, Julia W. Y. Kam, Michael S. Franklin, and Jonathan W. Schooler. (2012). Inspired by Distraction: Mind Wandering Facilitates Creative Incubation. "Psychological Science," 23(10), 1117-1122.

Bardeen. (2024). Find LinkedIn Hiring Managers: A Step-by-Step Guide. Retrieved from https://www.bardeen.ai/answers/how-to-find-the-hiring-manager-on-linkedin

Batdı, V. (2024). Evaluation of the effectiveness of critical thinking training on critical thinking skills and academic achievement by using bi-meta method. "Review of Education," 12(2).

Bezos, J., and Isaacson, W. (2021). "Invent and Wander: The Collected Writings of Jeff Bezos, With an Introduction by Walter Isaacson." Boston: Harvard Business Review Press.

Blinder, Alan. "College Financial Troubles Affect the Student Experience." New York Times, August 12, 2025.

"Boeing Is Paralyzed, and This Failing of Its Executives... Is to Blame." "MarketWatch," January 2024.

Breakstone, J., et al. (2021). Students' civic online reasoning: A national portrait. "Educational Researcher," 50(8), 505-515.

Buckner, R., Andrews-Hanna, J. R., and Schacter, D. L. (2008). The Brain's Default Network: Anatomy, Function, and Relevance to Disease. "Annals of the New York Academy of Sciences," 1124(1), 1-38.

Bureau of Labor Statistics. "Employment Projections: 2023-2033 Summary." U.S. Department of Labor, August 29, 2024. https://www.bls.gov/news.release/ecopro.nr0.htm.

Bureau of Labor Statistics. "Fastest Growing Occupations: 20 occupations with the highest projected percent change of employment between 2023–33." "Occupational Outlook Handbook." https://www.bls.gov/ooh/fastest-growing.htm.

Bureau of Labor Statistics. "Industry and occupational employment projections overview and highlights, 2023–33." "Monthly Labor Review," November 2024. https://doi.org/10.21916/mlr.2024.21.

Bureau of Labor Statistics. "New BLS employment projections: 3 charts." U.S. Department of Labor Blog, September 6, 2024. https://blog.dol.gov/2024/09/06/new-bls-employment-projections-3-charts.

Bureau of Labor Statistics. "Occupational employment projections to 2024." "Monthly Labor Review," December

2015. https://www.bls.gov/opub/mlr/2015/article/occupational-employment-projections-to-2024.htm.

"The Case Against Boeing." "The New Yorker," November 2019.

Christoff, Kalina, Alan M. Gordon, Jonathan Smallwood, Rachelle Smith, and Jonathan W. Schooler. (2009). Experience Sampling During fMRI Reveals Default Network and Executive System Contributions to Mind Wandering. "Proceedings of the National Academy of Sciences," 106(21), 8719-8724.

Clark, B. (2023). How to Research a Company for an Interview: 10 Steps. Career Sidekick. Retrieved from

Clear, J. (2018). "Atomic Habits: An Easy & Proven Way to Build Good Habits & Break Bad Ones." Avery.

Denning, Steve. "The Boeing Debacle: Seven Lessons Every CEO Must Learn." "Forbes," January 17, 2013.

Duhigg, C. (2012). "The Power of Habit: Why We Do What We Do in Life and Business." Random House.

Duke, A. (2018). "Thinking in Bets: Making Smarter Decisions When You Don't Have All the Facts." Portfolio.

Dweck, C. S. (2006). "Mindset: The New Psychology of Success." Random House.

Enago Academy. (2019). 9 Essential Things to Research Before a Job Interview.

Escobari, Marcela, et al. "Workforce of the Future initiative." Brookings Institution, 2018-2025. https://www.brookings.edu/collection/workforce-of-the-future-initiative/.

"Ex-OpenAI Board Member Reveals Why Sam Altman Was Briefly Ousted as CEO." "New York Post," May 29, 2024.

Facione, P. A. (1990). Critical thinking: A statement of expert consensus for purposes of educational assessment and instruction. The Delphi Report. California Academic Press.

Fast, Larry. "Boeing's Leadership Is at the Root Cause of Its Safety Failures." "EHS Today," July 16, 2019.

Fisher, R., & Ury, W. (1981). "Getting to Yes: Negotiating Agreement Without Giving In." Houghton Mifflin.

Flavell, J. H. (1979). Metacognition and cognitive monitoring: A new area of cognitive-developmental inquiry. "American Psychologist," 34(10), 906-911.

"For BlackBerry Leaders, Success Paved Way for Failure." "MoneyWatch," CBS News. Accessed July 2025.

Golden, Brighid. (2023). Enabling Critical Thinking Development in Higher Education through the Use of a Structured Planning Tool. "Irish Educational Studies" 42(4): 949–69.

Gonsalves, C. (2024). Generative AI's Impact on Critical Thinking: Revisiting Bloom's Taxonomy. "Journal of Marketing Education."

Haidt, J. (2012). "The Righteous Mind: Why Good People Are Divided by Politics and Religion." Pantheon Books.

Heath, C., & Heath, D. (2007). "Made to Stick: Why Some Ideas Survive and Others Die." Random House.

Heath, C., & Heath, D. (2013). "Decisive: How to Make Better Choices in Life and Work." Crown Business.

Heger, Brian. "Future of Jobs Report 2025 | World Economic Forum." BrianHeger.com, January 15, 2025. https://www.brianheger.com/future-of-jobs-report-2025-world-economic-forum/.

Hernandez, L. (2023). A Better Way to Message Hiring Managers on LinkedIn (with 6 message templates). LinkedIn.

Immordino-Yang, M. H., Christodoulou, J., and Singh, V. (2012). Rest Is Not Idleness: Implications of the Brain's Default Mode for Human Development and Education. "Perspectives on Psychological Science," 7(4), 352-364.

Janz, T., L. Hellervik, and D. C. Gilmore. Behavior Description Interviewing: New, Accurate, Cost-Effective. Boston: Allyn & Bacon, 1986

Jobscan. (2024). 19 Job Interview Tips That Will Get You Hired in 2025. Retrieved from https://www.jobscan.co/blog/job-interview-tips/

JobScore. (2025). Job Interview Statistics You Should Know in 2025. Retrieved from https://www.jobscore.com/articles/interviewing-statistics/

Kahneman, D. (2011). "Thinking, Fast and Slow." Farrar, Straus and Giroux.

Kahneman, D. and Tversky, A. "The Framing of Decisions and the Psychology of Choice," "Science" 211, no. 4481 (January 1981): 453-458.

Karikó, K., et al. (2005). Suppression of RNA recognition by Toll-like receptors: The impact of nucleoside modifications and the evolutionary origin of RNA. "Immunity," 23(2), 165-175.

Kaufman, Scott Barry, and Jerome L. Singer. (2016). The Origins of Positive-Constructive Daydreaming. In "The Wiley Handbook of Positive Clinical Psychology," edited by Alex M. Wood and Judith Johnson, 429-448. Hoboken, NJ: John Wiley & Sons.

Kelley, D. (2013). "The Art of Reasoning" (4th Edition). W. W. Norton & Company.

Klinger, Eric. (2010). Daydreaming and Fantasizing: Thought Flow and Motivation. In "Handbook of Individual Differences in Cognition," edited by Aleksandra Gruszka, Gerald Matthews, and Błażej Szymura, 225-239. New York: Springer.

LHH. (2025). How to Ask About Growth Opportunities During an Interview. Retrieved from https://www.lhh.com/us/en/insights/how-to-ask-about-growth-opportunities-during-an-interview/

Manyika, James, et al. "A new future of work: The race to deploy AI and raise skills in Europe and beyond." McKinsey Global Institute, May 21, 2024. https://www.mckinsey.com/mgi/our-research/a-new-future-of-work-the-race-to-deploy-ai-and-raise-skills-in-europe-and-beyond.

Manyika, James, et al. "Generative AI and the future of work in America." McKinsey Global Institute, July 26, 2023. https://www.mckinsey.com/mgi/our-research/generative-ai-and-the-future-of-work-in-america.

Manyika, James, et al. "Jobs lost, jobs gained: What the future of work will mean for jobs, skills, and wages." McKinsey Global Institute, November 28, 2017. https://www.mckinsey.com/featured-insights/future-of-work/jobs-lost-jobs-gained-what-the-future-of-work-will-mean-for-jobs-skills-and-wages.

Manyika, James, et al. "Skill shift: Automation and the future of the workforce." McKinsey Global Institute, May 23, 2018. https://www.mckinsey.com/featured-insights/future-of-work/skill-shift-automation-and-the-future-of-the-workforce.

Maslow, Abraham H. 1943. "A Theory of Human Motivation." Psychological Review 50 (4): 370-96

McInerny, D. Q. (2004). "Being Logical: A Guide to Good Thinking." Random House.

Minto, Barbara. The Pyramid Principle: Logic in Writing and Thinking. London: Minto International, 1990

Monster Career Advice. Do your research before a job interview. Retrieved from https://www.monster.com/career-advice/article/interview-company-research

Mooneyham, Benjamin W., and Jonathan W. Schooler. (2013). The Costs and Benefits of Mind-Wandering: A Review. "Canadian Journal of Experimental Psychology," 67(1), 11-18.

OECD. "OECD Employment Outlook 2024." OECD Publishing, July 2024. https://www.oecd.org/en/publications/oecd-employmen t-outlook-2024_ac8b3538-en.html.

Okolo, Chinasa T., et al. "Generative AI, the American worker, and the future of work." Brookings Institution, January 29, 2025. https://www.brookings.edu/articles/generative-ai-the-americ an-worker-and-the-future-of-work/.

Paul, R., & Elder, L. (2019). "The Miniature Guide to Critical Thinking: Concepts and Tools." Foundation for Critical Thinking Press.

Purdue University Global. How to Research a Company for a Job Interview. Retrieved from https://www.purdueglobal.edu/blog/car eers/research-company-job-interview/

Raichle, Marcus E., and Abraham Z. Snyder. (2007). A Default Mode of Brain Function: A Brief History of an Evolving Idea. "NeuroImage," 37(4), 1083-1090.

Rhone, Kailyn "Saving for College Once Felt Essential. Some Parents Are Rethinking Their Plans." "New York Times," July 26, 2025.

Ruiz-Rojas, L.I.; Salvador-Ullauri, L.; Acosta-Vargas, P. (2024) Collaborative Working and Critical Thinking: Adoption of Generative Artificial Intelligence Tools in Higher Education. "Sustainability," 16(13), 5367.

Schooler, Jonathan W., Erik D. Reichle, and David V. Halpern. (2004). Zoning Out While Reading: Evidence for Dissociations Between Experience and Metacognition. In "Thinking and Seeing: Visual Metacognition in Adults and Children," edited by Daniel T. Levin, 203-226. Cambridge, MA: MIT Press.

Sherpact. (2025). Job Market Analysis 2025: Top Trends and Career Strategies. Retrieved from https://sherpact.com/job-market-analysis/

Smallwood, Jonathan, and Jonathan W. Schooler. (2016). The Science of Mind Wandering: Empirically Navigating the Stream of Consciousness. "Annual Review of Psychology," 67, 487-518.

Sobel, Dave. "Inside OpenAI: The Controversy and Consequences of Sam Altman's Leadership Style." "The Business of Tech," December 13, 2023.

Sridharan, Mithun A. "SCRAP Framework: Specific, Concise, Relevant, Accurate, Professional." Think Insights. Accessed August 5, 2025. (Founded and authored by Mithun A. Sridharan.)

Stirone, Shannon. "How NASA Engineered Its Own Decline." "The Atlantic," September 20, 2025.

Tetlock, P., & Gardner, D. (2015). "Superforecasting: The Art and Science of Prediction." Crown Publishers.

Thaler, R. H., & Sunstein, C. R. (2008). "Nudge: Improving Decisions About Health, Wealth, and Happiness." Yale University Press.

"Theranos." Wikipedia contributors. "Wikipedia." Last modified May 2025.

Thompson, Derek. "The OpenAI Mess Is About One Big Thing." "The Atlantic," November 22, 2023.

Tiku, Nitasha. "Warning from OpenAI Leaders Helped Trigger Sam Altman's Ouster." "The Washington Post," December 8, 2023.

Toulmin, S. E. (2003). "The Uses of Argument" (Updated Edition). Cambridge University Press.

Tversky, A., & Kahneman, D. (1974). Judgment under uncertainty: Heuristics and biases. "Science," 185(4157), 1124-1131.

"United States presidential election of 1948." Britannica. Accessed August 3, 2025. https://www.britannica.com/event/United-States -presidential-election-of-1948

U.S. Bureau of Labor Statistics. "Occupational Outlook Handbook." Washington, DC: U.S. Bureau of Labor Statistics, 2024. ht tps://www.bls.gov/ooh/.

van Eemeren, F. H., & Grootendorst, R. (2004). "A Systematic Theory of Argumentation." Cambridge University Press.

Walton, D. (2008). "Informal Logic: A Pragmatic Approach" (2nd Edition). Cambridge University Press.

Welch, Suzy. 10-10-10: A Life-Transforming Idea. New York: Scribner, April 14, 2009

White, J., Q. Fu, S. Hays, M. Sandborn, C. Olea, H. Gilbert, A. Elnashar, J. Spencer-Smith, and D. C. Schmidt. "A Prompt Pattern Catalog to Enhance Prompt Engineering with ChatGPT." arXiv, February 21, 2023

Wineburg, S., & McGrew, S. (2019). Lateral reading and the nature of expertise: Reading less and learning more when evaluating digital information. "Teachers College Record," 121(11), 1-40.

World Economic Forum. "Future of Jobs Report 2025: 78 Million New Job Opportunities by 2030 but Urgent Upskilling Needed to Prepare Workforces." Press release, January 8, 2025. https://www.weforum.org/press/2025/01/future-of-jobs-report-202 5-78-million-new-job-opportunities-by-2030-but-urgent-upskilling -needed-to-prepare-workforces/.

World Economic Forum. "Future of Jobs Report 2025: The jobs of the future – and the skills you need to get them." January 2025. https://www.weforum.org/stories/2025/01/future-of-jobs-re port-2025-jobs-of-the-future-and-the-skills-you-need-to-get-them/.

World Economic Forum. "Future of Jobs Report 2025: What's shaping the future of the global workforce?" January 2025. https://www.weforum.org/stories/2025/01/future-of-jobs-report-2025-whats-shaping-the-future-of-the-global-workforce/.

World Economic Forum. "The Future of Jobs Report 2025." World Economic Forum, January 2025. https://www.weforum.org/publications/the-future-of-jobs-report-2025/.

Xie, Yan, Jo Smith, and Maree Davies. (2025). The Evolution of Critical Thinking in Chinese Education Context: Policy and Curriculum Perspectives. "International Studies in Sociology of Education," February, 1–24.

QUICK REFERENCE GUIDE

<u>Critical Thinking Checklists</u>

Keep this guide handy because it will help you in making determinations about many things in your life. You don't have to remember everything verbatim from this book, and this is simply a small bit of help.

First, before you begin going over this checklist, I want you to do something for yourself. What is it? Please go at your own pace and don't think that you have to run like some sprinter in a race to get finished. That's not the point. The point is that you should leisurely go through this material and work on it as you can. In fact, in some places it might be best for you to jot down what you are being asked to do and consider it for a few hours or even a day. In order to get the most from the material, you have to permit it to percolate a bit, and that takes thinking and consideration, not racing to finish.

Throughout the book, you probably have found areas that you would like to stop and spend more time on. That's fine. As I've always said throughout the book and I'm repeating now, go at your own pace. This is a book that you can come back to in any section, revisit it and rework what it is asking you to do. Allow yourself time to think about the examples it's asking you to provide in specific instances.

Don't just come up with the first thing that comes to mind. Take a little time. As you can see, number one below tells you not to proceed quickly. If there's one thing I want to underscore, that's it. Go slow and be like the turtle and the hare in that famous race where the turtle actually wins.

1. Don't feel that you have to run through this book or this checklist quickly. That's not the point of any of this.

2. What I really want you to do is take your time.

In order to help you utilize this checklist most efficiently, I'd like to make several points here that are particularly relevant:

Before Accepting Any Claim

• Who is making this claim, and what might motivate them? Don't be hesitant to ask.

   • What evidence is provided? Where did they find that?

   • How reliable is this evidence?

   • What alternative explanations exist? How else could that be explained?

   • What would I need to see to be more confident this is true?

Before Making Important Decisions

- What are my options?
  - What are the likely consequences of each option?
  - What am I uncertain about?
  - What additional information would be helpful?
  - What are my underlying values and priorities?
  - How will I know if this decision was right?

Before Engaging in Debate or Discussion

- Do I understand the other person's position?
  - What do we agree on?
  - What evidence would change my mind?
  - Am I trying to understand or just to win?
  - What might I be missing?

Cost-Benefit Analysis

1. List all costs (financial, time, opportunity, risks).
   2. List all benefits (immediate and long-term).
   3. Consider the probability and timing of each.
   4. Compare total expected costs to total expected benefits.

Decision Matrix

1. List your options.

   2. List your criteria.

   3. Weight criteria by importance.

   4. Rate each option on each criterion.

   5. Calculate weighted scores.

Scenario Planning

1. Best case: What if everything goes well?

   2. Worst case: What if things go badly?

   3. Most likely: What's the realistic outcome?

   4. Wild card: What unexpected events might occur?

   5. Plan for multiple scenarios.

Question for Different Contexts

For Evaluating Information

- What evidence supports this claim?
  - Who benefits from me believing this?
  - What would someone who disagrees say?
  - How could I verify this information?

For Problem-Solving

- What's the real problem here?

- What are we assuming?
- What would this look like if it were easy?
- Who else has solved similar problems?

For Decision-Making

- What would I choose if I had perfect information?
  - What would I advise a friend in this situation?
  - What would I think about this decision in five years?
  - What's the worst thing that could happen if I'm wrong?

Information Red Flags

- Extreme emotional language.
  - Claims that seem too good/bad to be true.
  - Vague attribution ("experts say" and "studies show"). Which ones?
  - Pressure to share immediately.
  - Attacks on people rather than ideas.

Decision-Making Red Flags

- Making important decisions when highly emotional.
  - Failing to consider alternatives.
  - Ignoring potential negative consequences.
  - Deciding based on what others expect rather than your own values.
  - Rushing decisions that don't need to be rushed.

Discussion Red Flags

• Personal attacks instead of addressing arguments.

• Unwillingness to acknowledge any valid points from the other side.

• Changing the subject when challenged.

• Claiming certainty about uncertain things.

• Refusing to specify what evidence would change your mind.

When You Need to Decide Quickly

1. Pause and take three deep breaths. It's OK to take a short time out.

2. Ask, "What would I advise someone else in this situation?"

3. Consider, "What's the worst reasonable outcome of each option?"

4. Choose based on your core values.

5. Decide when you'll reassess, if possible.

When Confronted with Suspicious Information

1. Ask, "Who benefits if I believe this?"

2. Check: "Does this align with what reliable sources have said?"

3. Ask yourself, "Am I being required to take immediate action?"

4. If in doubt, wait and verify later.

When Emotions Are High

1. Acknowledge the emotion: "I'm feeling angry/scared/excited about this."

    2. Ask, "How might this emotion be affecting my thinking?"

    3. Consider, "What would I think about this tomorrow?"

    4. If possible, delay important decisions until emotions settle

About the Author

D r. Patricia A. Farrell is a licensed psychologist, published author of multiple self-help books and videos, former WebMD psychologist expert/consultant, medical consultant for Social Security Disability Determinations, Alzheimer's psychiatric researcher at Mt. Sinai Medical Center (NYC), and an educator who has taught at the college, graduate, and postgraduate levels.

Her influence extends to the pharmaceutical and marketing industries, where she serves as a consultant and has appeared on major TV news programs in the US and abroad. In addition, Dr. Farrell provides continuing education modules for mental healthcare professionals and has contributed to USMLE medical school prep courses. She shares her knowledge through her YouTube channel and her daily contributions to **Bluesky** (@carpenter22,bsky.social) and Medium. com articles. Dr. Farrell's achievements are recognized in *Who's Who in the World, Who's Who in America,* and *Who's Who in American Women.*

A member of the American Psychological Association and the SAG-AFTRA union, Dr. Farrell is a former board member of the NJ Board of Psychological Examiners, a former psychiatry preceptor at UMDNJ, and a former board of directors member of Bergen Pines Hospital (now Bergen Regional Hospital).

Books by Patricia A. Farrell, Ph.D.

How to Be Your Own Therapist

When You Can't Pour From an Empty Glass: CBT Skills for Exhausted Caregivers

It's Not All in Your Head: Anxiety, Depression, Mood Swings and Multiple Sclerosis

Unfiltered: Beneath the noise of our thoughts lies the true narrative of our minds

Unfiltered Again: A behind-the-scenes look at healthcare, medicine and mental health

When You Can't Pour From an Empty Glass: CBT Skills for Exhausted Caregivers

A Social Security Disability Psychological Claims Handbook: A simple guide to understanding your SSD claim for psychological impairments and unraveling the maze of decision-making

A Social Security Disability Psychological Claims Guidebook for Children's Benefits

The Disability Accessible US Parks in All 50 States: A Comprehensive Guide

Birding in the US NOW!: A birding guide for individuals with disabilities

A Special Request

If this book has touched your heart, sparked your curiosity, or simply entertained you along the way, I'd be incredibly grateful if you could take a moment to share your thoughts with a review on Amazon or wherever you discovered this book. Your words not only help other readers find books they'll love, but they also mean the world to authors like me who pour their hearts into every page. Thank you for being part of this journey, and for helping stories find their way to the readers who need them most.